Advances in Intelligent Systems and Computing

Volume 844

Series editor

Janusz Kacprzyk, Polish Academy of Sciences, Warsaw, Poland
e-mail: kacprzyk@ibspan.waw.pl

The series "Advances in Intelligent Systems and Computing" contains publications on theory, applications, and design methods of Intelligent Systems and Intelligent Computing. Virtually all disciplines such as engineering, natural sciences, computer and information science, ICT, economics, business, e-commerce, environment, healthcare, life science are covered. The list of topics spans all the areas of modern intelligent systems and computing such as: computational intelligence, soft computing including neural networks, fuzzy systems, evolutionary computing and the fusion of these paradigms, social intelligence, ambient intelligence, computational neuroscience, artificial life, virtual worlds and society, cognitive science and systems, Perception and Vision, DNA and immune based systems, self-organizing and adaptive systems, e-Learning and teaching, human-centered and human-centric computing, recommender systems, intelligent control, robotics and mechatronics including human-machine teaming, knowledge-based paradigms, learning paradigms, machine ethics, intelligent data analysis, knowledge management, intelligent agents, intelligent decision making and support, intelligent network security, trust management, interactive entertainment, Web intelligence and multimedia.

The publications within "Advances in Intelligent Systems and Computing" are primarily proceedings of important conferences, symposia and congresses. They cover significant recent developments in the field, both of a foundational and applicable character. An important characteristic feature of the series is the short publication time and world-wide distribution. This permits a rapid and broad dissemination of research results.

More information about this series at http://www.springer.com/series/11156

Grzegorz Sierpiński
Editor

Integration as Solution for Advanced Smart Urban Transport Systems

15th Scientific and Technical Conference
"Transport Systems. Theory & Practice 2018"
Selected Papers

 Springer

Editor
Grzegorz Sierpiński
Faculty of Transport
Silesian University of Technology
Katowice, Poland

ISSN 2194-5357 ISSN 2194-5365 (electronic)
Advances in Intelligent Systems and Computing
ISBN 978-3-319-99476-5 ISBN 978-3-319-99477-2 (eBook)
https://doi.org/10.1007/978-3-319-99477-2

Library of Congress Control Number: 2018951635

This Springer imprint is published by the registered company Springer Nature Switzerland AG
The registered company address is: Gewerbestrasse 11, 6330 Cham, Switzerland

Preface

Transport systems constantly require development, which may include among others infrastructural, organizational or legal character. Corrective actions implemented on the macroscale in smart cities are targeted mainly for the implementation of an efficient, cost-effective, safe and environmentally friendly transport. One of the directions of development of modern transport systems is integration, which supports both the transport of people and loads. Advanced intelligent communication systems make it easy to change the means of transport during a trip planning process. Similarly, advanced and complex models provide support for calculation and optimization of supply chains, also according to the pro-ecological criterion.

This publication contains selected papers submitted to and presented at the 15th Scientific and Technical Conference "Transport Systems. Theory and Practice" (TSTP2018) organized by the Department of Transport Systems and Traffic Engineering at the Faculty of Transport of the Silesian University of Technology in Katowice, Poland. The problems addressed in the book entitled *Integration as Solution for Advanced Smart Urban Transport Systems* have been divided into four parts:

- Part 1. The latest achievements of intelligent transport systems,
- Part 2. Advanced data collecting, analysis and intelligent support for decision-making,
- Part 3. Modelling, optimization and evaluation in smart transport,
- Part 4. Integrated urban passenger and freight transport.

The articles included in the publication are expressions of case study-based scientific and practical approach to the problems of contemporary transport systems. The proposed methods and models enable a system approach to assess current solutions. In turn, implementation proposals may support the improvement of the integrity of individual elements of transport systems, and thus increase its effectiveness on the global scale.

I would like to express my deepest gratitude to all authors, for reflecting the key problems of contemporary transport systems in a concise manner, as well as to reviewers, in recognition of their insightful remarks and suggestions without which this collection of papers would have never been published.

September 2018

Grzegorz Sierpiński

Organization

15th Scientific and Technical Conference "Transport Systems. Theory and Practice" (TSTP2018) is organized by the Department of Transport Systems and Traffic Engineering, Faculty of Transport, Silesian University of Technology, Poland.

Organizing Committee

Organizing Chair

Grzegorz Sierpiński — Silesian University of Technology, Poland

Members

Renata Żochowska
Grzegorz Karoń
Krzysztof Krawiec
Aleksander Sobota
Marcin Staniek
Ireneusz Celiński

Barbara Borówka
Kazimierz Dąbała
Marcin J. Kłos
Damian Lach
Piotr Soczówka

The Conference Took Place Under the Honorary Patronage

Ministry of Infrastructure
Marshal of the Silesian Voivodeship
Silesian Voivode

Scientific Committee

Stanisław Krawiec (Chairman)	Silesian University of Technology, Poland
Rahmi Akçelik	SIDRA SOLUTIONS, Australia
Tomasz Ambroziak	Warsaw University of Technology, Poland
Henryk Bałuch	The Railway Institute, Poland
Roman Bańczyk	Voivodeship Centre of Road Traffic in Katowice, Poland
Werner Brilon	Ruhr-University Bochum, Germany
Margarida Coelho	University of Aveiro, Portugal
Boris Davydov	Far Eastern State Transport University, Khabarovsk, Russia
Mehmet Dikmen	Baskent University, Turkey
Domokos Esztergár-Kiss	Budapest University of Technology and Economics, Hungary
József Gál	University of Szeged, Hungary
Andrzej S. Grzelakowski	Gdynia Maritime University, Poland
Mehmet Serdar Güzel	Ankara University, Turkey
Józef Hansel	AGH University of Science and Technology Cracow, Poland
Libor Ižvolt	University of Žilina, Slovakia
Marianna Jacyna	Warsaw University of Technology, Poland
Nan Kang	Tokyo University of Science, Japan
Jan Kempa	University of Technology and Life Sciences in Bydgoszcz, Poland
Michael Koniordos	Pireaus University of Applied Sciences, Greece
Bogusław Łazarz	Silesian University of Technology, Poland
Zbigniew Łukasik	Kazimierz Pulaski University of Technology and Humanities in Radom, Poland
Michal Maciejewski	Technical University Berlin, Germany
Elżbieta Macioszek	Silesian University of Technology, Poland
Ján Mandula	Technical University of Košice, Slovakia
Sylwester Markusik	Silesian University of Technology, Poland
Antonio Masegosa	IKERBASQUE Research Fellow at University of Deusto Bilbao, Spain
Agnieszka Merkisz-Guranowska	Poznań University of Technology, Poland

Elżbieta Załoga   University of Szczecin, Poland
Stanisława Zamkowska   Blessed Priest Władysław Findysz University
                              of Podkarpacie in Jaslo, Poland
Jacek Żak   Poznań University of Technology, Poland
Jolanta Żak   Warsaw University of Technology, Poland

Referees

Rahmi Akçelik
Marek Bauer
Przemysław Borkowski
Werner Brilon
Margarida Coelho
Piotr Czech
Domokos Esztergár-Kiss
Michal Fabian
Barbara Galińska
Róbert Grega
Mehmet Serdar Güzel
Katarzyna Hebel
Nan Kang

Peter Kaššay
Jozef Kuĺka
Michał Maciejewski
Elżbieta Macioszek
Krzysztof Małecki
Martin Mantič
Silvia Medvecká-Beňová
Katarzyna Nosal Hoy
Romanika Okraszewska
Asier Perallos
Hrvoje Pilko
Antonio Pratelli
Michal Puškár

Piotr Rosik
Alžbeta Sapietová
Grzegorz Sierpiński
Marcin Staniek
Dariusz Tłoczyński
Andrzej Więckowski
David Williams
Grzegorz Wojnar
Adam Wolski
Ninoslav Zuber

Contents

Integrated Urban Passenger and Freight Transport

The Latest Achievements of Intelligent Transport Systems

Basic Framework for the Energy-Effective Train Dispatching

Vadim Gopkalo[1]([⊠]), Boris Davydov[2], and Alexandr Godyaev[2]

[1] Institute of Natural Sciences, Far Eastern State Transport University,
Khabarovsk, Russian Federation
vng@yandex.ru
[2] Institute of Control, Automation and Telecommunications,
Far Eastern State Transport University, Khabarovsk, Russian Federation
dbi@rambler.ru

Abstract. The problem of modeling and optimal control the train flow is considered in relation to the mixed passenger and freight traffic. The paper describes main principles used to choose traffic adjustments and to evaluate change of the consumption value due to an operative control activity. Well-defined and rapid alteration of the train speed trajectory due to impediment arise is the basic element of traffic operative adjustments. The paper proposes the methodology of calculating the travel mode parameters in case of disturbance and the optimal distribution of time margin added to trajectory elements. Correlation analysis application of the real data allows revealing the regularities in normal and disturbed train run.

Keywords: Train traffic · Online adjustments · Energy consumption
Modeling · Optimal decision making

1 Introduction

Adjustments of train traffic are made in order to eliminate deviations that have already occurred or are possible in the near future. These deviations from the schedule leads to the train lateness and to overruns of energy and money. You must have a high-quality dispatching to reduce the level of losses. Such an adaptive train management involves elaboration of the optimal schedule adjustments as well as the precise and timely its implementation.

Most of the current literature explores the problem of finding the rescheduling solutions which are optimized according to the criterion of punctuality the train traffic. This approach is valid for the control of passenger trains and of freight services are moving according to exact schedule. It is often used operational adjustments which significantly increase traction energy consumption. There are a few papers that examine the problem of reducing energy consumption while maintaining a high level of punctuality. Obviously, this problem needs an additional study.

There is an intense flow of mixed passenger and freight trains at the numerous conventional mainlines in Russia, China, the United States and the other countries. It is known that requirement on the accurate schedule execution when driving freight train

© Springer Nature Switzerland AG 2019
G. Sierpiński (Ed.): TSTP 2018, AISC 844, pp. 3–12, 2019.
https://doi.org/10.1007/978-3-319-99477-2_1

is much weaker than in the traveling of passenger train. Schedule of ordinary freight trains contains large amount of the time supplements which compensate the random perturbations. Time reserve appears in periods when there is a little of random obstacles to the train traffic. This supplement should be used to save energy.

Modeling of train traffic is the most important part of operational rescheduling process. The major research in the field of traffic modeling is devoted to solving the problem of train delay minimizing. There is a lack of models that predict deployment of the train traffic situation and determine the optimal adjustment based on energy criterion. The presenting paper is devoted to solving the issues that allows eliminating this drawback.

The principles underlying the dispatching adjustments are provided in Sect. 3 after the analytical literature review (Sect. 2). In the last section we present a brief description of methods for operational traffic management from the perspective of traction energy consumption reducing.

The requirement for precise execution the schedule of freight trains is much weaker than for the travel of passenger trains. Large time margins that compensate the random disturbances are included to the timetable of ordinary freight trains. When there is a small number of an obstacle to the movement of the train, then there is a time reserve. This reserve should be used to save a power resource. The Sect. 4 describes the approach of energy consumption reduced by making optimal distribution of the time margin which is included into the schedule. Used framework is effective in the case of unimpeded train traffic is expected. The final Sect. 5 carries detailed analysis of dynamics the energy consumption. Fast dynamics of traction energy is studied in order to obtain the regularities that are used for the consumption volume forecast.

2 Literature Review

Most of the published work on the problem of operative train traffic rescheduling is based on deterministic models which describe functioning the railway section. The paper [1] provides an extensive overview of recovery models and algorithms for real-time railway rescheduling. The paper deals with the problem of passenger orientation when searching for the optimal traffic adjustments. The main types of problems that are solved in the process of decision making are conflict detection and resolution (CDR) and train speed coordination (TSC) [2–5]. In these problems, the best order of train traffic determined to eliminate deviations from the schedule (lateness) which leads to conflicts arise. Most of the works explores the algorithms that make corrections of the passenger train traffic on busy line. Typically, it is considered the total value of delay of all trains in a given time period. Authors that attempt to use economic indicators of passing quality the passenger flow, one way or another interprets the quantity of delays. The decisions for rescheduling which minimize delays arc calculated by using the discrete models of traffic and the integer or partially integer linear programming [for example, 6, 7].

Local punctuality in the freight flow movement recedes into the background. Here you want to use other criteria which adequate to economic interests of the railway operator company. The analysis shows that the operating costs for passing trains may be used as a criterion that corresponds to the freight trains traffic on the section. The same applies to the part of profit the railway operator. Formulation of such the economic criterion is made in [8].

Many papers have been published [e.g. 10, 11] which address to the creation the trajectory of the train determining a minimum energy consumption. Rational allocation of the extra time between the elementary path segments is provided by the technique of timetable adjustment that aims the fuel economy. Only a small number of works devoted to solving the problem of reducing both the energy consumption and the costs through an efficient operational traffic management. As an example we present the paper [9] the authors of which are exploring this problem for suburban passenger traffic. This problem is more fully developed in the paper [11]. The proposed approach combines real-time rescheduling (performed after an incident) with train punctuality facilitated by providing dynamic schedule information to train drivers. This type of dispatching is very suitable for consumption reducing.

The research group [12] develop fuel consumption model for typical traffic facilities using artificial neural networks. This model allows combining microscopic and macroscopic analyses of consumption. The macro-model was developed based on an aggregation of information obtained at the microscopic level. This approach seems to be effective and can be used in forecasting the consumption of rail traffic.

The presented brief overview shows the absence of papers that investigate the soft dynamics of energy consumption. This raises the need to study this problem.

3 Preventive and Reactive Adjustments of the Schedule Disturbed

We initially believe that train traffic is continuously undergoes to random disturbing factors. Time margins are included in the schedule to reduce the influence of these perturbations. The additives weakly expands a run times and headways not to reduce the railway capacity. The adaptive traffic control allows overcoming this contradiction between the accuracy of the schedule and the capacity decrease.

There are two strategies for the operative adjustment of train traffic which serves to reduce delays in the occurrence of disturbances [8]. The first one is immediate changing the speed mode of a specific train when a signal about an obstacle at its path is received. This maneuver is performed irrespective of approaching distance to the point of possible conflict. Obviously, it is necessary to provide a reliable and quick transmission of signal about the event to the locomotive by radio. The second type of strategies involves the preventive change of some part the schedule when such an intervention does not lead to a failure in train traffic. This type of adjustment is applied in periods when there is a decline in traffic intensity and there are no serious failures of an open track and station infrastructures. In this period, the train manager purposefully assigns the group of trains that has a longer run time that is the economical traffic mode. This mode is most suitable for a freight train travel. Increased run-time reserve allows using the effective techniques of locomotive driving and provides significant energy savings. The example of combination the economical and the standard (intensive) modes is illustrated in Fig. 1.

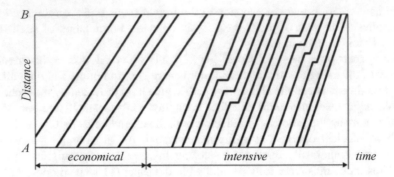

Fig. 1. Fragment of schedule with the economical and intensive train traffic

We consider the methodology for analyzing the features of train flow and rational choice of a traffic mode as *the on-line integral flow adjustments*. The paper [13] uses the stochastic model which allows determining the local zone of schedule space in which it is expedient to introduce the economical traffic mode. The planned mode of motion without stopping frequently interrupted perturbing random obstacles, therefore, when analyzing the amount of energy we have to divide the sections of the route without stops and stopovers.

Operational adjustment of speed profile the single train is main adjustment option which aims to reduce energy consumption in the case of unscheduled stop. Here we believe the main type of maneuver that is carried out to prevent unscheduled stop is timely speed reduction of following train (see Fig. 2). Train 2 resets the speed at the point A when receiving information on short stopping the preceding train. The reduced speed value is maintained until the train 1 will get out at the safe interval Δs_{sch} that is prescribed by the schedule. Herewith you need to solve the following problem. It is necessary to choose such a speed of the train 2 that it not to get into the zone of impact the yellow aspect after receiving a signal about arise of stop the train 1. This is achieved when current value of the headway Δs at closest approach of both trains exceeds the overall length Δs_{app} of two block sections.

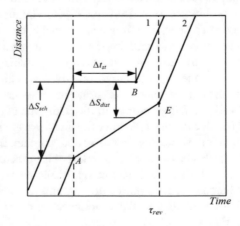

Fig. 2. Fragment of adjusted schedule in case of unplanned stop

If the headway is equal to $\Delta s_{sch} = v\,\Delta t_{sch}$ (where v is the planned speed) then complied the following condition:$\Delta s > \Delta s_{dist}$. This condition is satisfied when you are set the certain train speed value:

$$v_2' < \frac{\Delta s_{dist}}{\Delta t_{st}}\left(\frac{\Delta s_{sch}}{\Delta s_{dist}} - 1\right) \tag{1}$$

During a slow traveling mode, the driver can apply the optimal speed trajectory. This would result in savings of amount at least 50 kWh when moving passenger train and more than 100 kWh for heavy freight train traveling. Increasing the travel time involves the use of additional energy-saving techniques. These techniques include: the full use of inertia of the train and ensuring a rational mode of escape before stops.

4 Optimal Distribution of Train Travel Time by Path Elements

In the train schedule there are time margins that serve to eliminate the influence of delays. These additives can be used to reduce energy consumption in the case that unimpeded train traffic is expected. The methodology of time distribution is that the travel times increase on those elements of the path where there is greater sensitivity

$$S_{Ai} = \frac{\partial A_i(T_{Xi})}{\partial T_{Xi}} \tag{2}$$

to these changes.

The problem of ensuring the optimal control of a train movement is formalized as follows. Suppose there is energy function of n variables

$$A_S = f\left(T_{X0}, T_{X1}, \ldots, T_{X(n-1)}\right)$$

predetermined by the expression

$$A_S = \sum_i A_i(T_{Xi}) \tag{3}$$

It is required to find the values of variables

$$T_{X0}, T_{X1}, \ldots, T_{X(n-1)}$$

which lead to minimizing the function A_S. The condition for existence the minimum is determined by introducing an indefinite Lagrange multiplier:

$$\frac{\partial A_i(T_{Xi})}{\partial T_{Xi}} + \lambda = 0, \quad i = 0, 1, \ldots (n-1)$$

In the simplest case when

$$A_i(T_{Xi}) = K_i/T_{Xi},$$

the values of the run times are calculated by the formula:

$$T_{Xi} = \beta_i T_{y4},$$

where

$$\beta_i = \frac{\sqrt{K_i}}{\sum_i \sqrt{K_i}} I$$

K_i is the consumption sensitivity value to changes of the travel time at a section.

Note that a similar result was obtained in determining the most advantageous train traffic mode by the criterion of minimum operating costs.

In some cases, the function $A_i(T_{Xi})$ becomes more complex. Therefore, the system of equations must be solved by numerical methods, in particular by the method of dynamic programming. The basic functional equation of dynamic programming is written in the following form:

$$\min[A_S(T_S - \Delta T)] = \min[A(T_S)] + \min[\Delta A_i(k_i)],$$

where ΔT is the sampling interval, T_S is the scheduled travel time at the site and k_i is number of an iterative step.

The algorithm for determining the optimal time margin distribution contains $(k - 1)\, n$ calculation cycles for n interstation sites. This algorithm is small time consuming so it can be used in the onboard locomotive controller.

Experimental studies of effect the train traffic mode on energy consumption have been carried out at the Russian rail mainlines. The results of the research show, reduction of the energy consumption ΔA_i reaches 25 kWh for every minute of the run time increase when the long-distance passenger train is traveled. This is observed in areas with hilly and mountainous relief where it's efficient using the energy-saving techniques. A similar index for a freight train has a larger rate reaching 60–80 kWh.

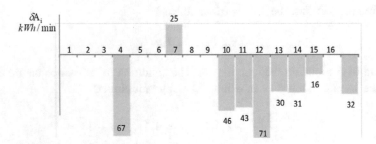

Fig. 3. Change in consumption with the run time increase by 1 min.

On the section of 300 km with a hard relief, there are eight interstations on which the rise of run time is effective, as you see on the Fig. 3. The optimization calculation shows, the increase the travel time of a freight train by 12.5% and rational use of the additional time gives an energy saving of 1570 kWh or 16% of the total consumption at the investigated section.

5 Study on the Energy Consumption Dynamics

We have investigated in detail the energy consumption at the electrified main line. Data on the current energy consumption, the coordinate and train speed were obtained from the onboard recorder. The volume of energy was determined for a train travel at elementary sections with different relief characteristic. One of them (type I) has the-heavy relief with lengthygrade. Other sections have a more smooth behavior (types II and III). Measurements were carried out on a large number of a train runs the same type. Measured values of power consumption for different sites are summarized in Table 1.

Table 1. Power consumption according to heaviness the site relief, kW * h

Site tipe	I	II	III
Length, km	15	40	20
Min volume	900	1207	641
Average volume	1112	1559	845
Max volume	1264	2055	1184

The analysis carried out in [14] showed that the scattering of a random variable characterizing run time is well approximated by the gamma-distribution, i.e.

$$g(t; s_i) = t^{\alpha-1} e^{-t/\beta} \frac{1}{\beta^{\alpha} \Gamma(\alpha)}$$

where $\alpha > 0$, $\beta > 0$ are some parameters, $\Gamma(\cdot)$ is the Euler gamma-function. The corresponding random variable Z has an expectation $\mathbf{E}Z = \alpha\beta$ and a variance $\mathbf{D}Z = \alpha\beta^2$. The mode of this distribution is equal to m = $(\alpha - 1)\beta$.

Studies carried out on the real railway section confirmed, the distribution of driving times the freight services becomes more asymmetrical as the train travels along the route. Figure 4 shows the energy histograms corresponding to some elementary rail sections and imposed on them gamma-densities $g(t; s_i)$.

Kolmogorov test shows, the samples 3 and 4 are more closely corresponds to gamma-distribution (with significance level of $\alpha = 0, 1$) than consumption scattering at the previous sections. Calculated distribution parameters are depicted in the Table 2.

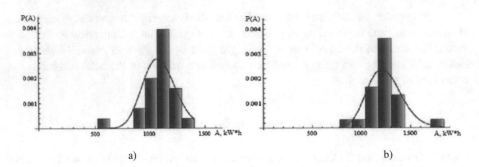

a)   b)

Fig. 4. Energy consumption distributions at the element of site: (a) initial element; (b) further element

Table 2. Distribution parameters of consumption at the different elementary sections, kWh

	1	2	3	4
Mean	1094	149	53.5	20.4
Variance	449	106	12.8	25.2
α	1240	171	37.0	28.1
β	579	241	3.2	135.9

Experimental data show that there is a sufficiently strong correlation of energy consumption and travel time over many elements of the site. The correlation index reaches values of 0.5–0.6. This is due to the presence of unplanned delays that increase both travel time and consumption. The statistical relationship is weakened when some experienced drivers use the time reserve for the energy efficient train control. Similar regularities are also observed in the traffic process of passenger trains.

Actualtrain speed value depends on current condition of the infrastructure and on quality the train control that implements the driver. It is necessary to separate these two impacts during analysis fulfillment to create a correct prediction of consumption, taking into account the likely future deviations from the schedule.

We propose original methodology to check the root of occurrence the energy overruns at the element of train path. This approach allows you to identify the influence of the disturbing factor which leads to exceeding the normative consumption. The factor mentioned is a deviation from standard trajectory as a result of driver's nonrational action or presence an obstacle to train traveling.

The developed approach is based on obtaining data on the traction energy and the implementation of its correlation analysis. Basic regularities determined most fully if you are considering the correlation of energy consumption of given train unit with the characteristic of the path relief and with energy function other trains. The values of correlation index calculated for the consumption and the path relief are depicted at Fig. 5.

The data presented at the picture are showing the following regularities. The correlation index rises with increasing severity of the site relief. This is due to the fact, on a site that includes a steep and extended rise there is a very narrow spectrum of options for implementation the traction trajectory.

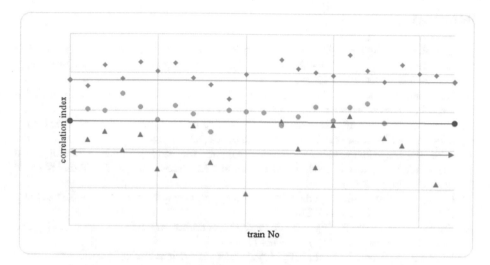

Fig. 5. Correlation index of the energy consumption and the path relief

Drivers use various run trajectories on the light sections so the characteristics of energy consumption by different trains correlate poorly. This explains the great difference in the amount of energy that is spent by the experienced and inexperienced drivers when they have freedom of action during the locomotive control.

The calculated correlation index of the consumption and of the vertical profile of the path is weakened when unscheduled stops arise. Thus, the presence of short stops when train's 22, 23 run at the section I reduces the correlation index from 0.8 to 0.5.

6 Conclusions

Optimum dispatching can significantly reduce the consumption of traction energy. It's necessary to implement the preventive traffic adjustments for energy reducing. Preventive dispatch control is based on the permanent monitoring the train coordinates and determining the optimal traffic mode. The economical mode assumes the optimal use of time reserves that arise when the intensity of the train flow decreases. Local adjustments provide the target train trajectory correction when the unscheduled delay arises. The result of this activity is a significant reduction in consumption mostly by the freight train travel. The proposed methodology of correlation analysis of traction energy allows you to carefully study the effect of various factors on resource overruns. Results obtained can improve the consumption prognosis and provide a tool to assess effective operational dispatch adjustments.

The research is partially funded by the JSC Russian Railways (grant 2016 for development of the scientific school).

References

1. Cacchiani, V., Huisman, D., Kidd, M., Kroon, L., Toth, P., Veelenturf, L., Wagenaar, J.: Overview of recovery models and algorithms for real-time railway rescheduling. Transp. Res. B **63**, 15–37 (2014)
2. D'Ariano, A., Pacciarelli, D., Pranzo, M.: A branch and bound algorithm for scheduling trains in a railway network. Eur. J. Oper. Res. **183**(2), 643–665 (2007)
3. Quan, L., Dessouky, M., Leachman, R.C.: Modeling train movements through complex rail networks. ACM Trans. Model. Comput. Simul. **14**, 32–76 (2004)
4. Törnquist, J.: Computer-based decision support for railway traffic scheduling and dispatching: a review of models and algorithms. In: Proceedings of ATMOS, Palma de Mallorca, Spain (2005)
5. Carey, M., Crawford, I.: Scheduling trains on a network of busy complex stations. Transp. Res. B **41**, 159–178 (2007)
6. Cordeau, J.-F., Toth, P., Vigo, D.: A survey of optimization models for train routing and scheduling. Transp. Sci. **32**(4), 380–404 (1998)
7. Huisman, D., Kroon, L., Lentink, R., Vromans, M.: Operations research in passenger railway transportation. Erasmus Research Institute of Management (ERIM). Research paper no. ERS-2005-023-LIS (2005)
8. Davydov, B., Dynkin, B., Chebotarev, V.: Optimal rescheduling for the mixed passenger and freight line. In: WIT Transactions on the Built Environment/14th International Conference on Railway Engineering Design and Optimization - COMPRAIL 2014, Rome, Italy, vol. 135, pp. 649–661 (2014). https://doi.org/10.2495/CR140541
9. Kuznetsova, A.: Optimization of technology of driving freight trains on stages on the criterion of minimum operating costs for mileage. Novosibirsk (2006). Кузнецова, А.А.: Оптимизация технологии вождения грузовых поездов по перегонам по критерию минимума эксплуатационных расходов по пробегу. Дисс…к.т.н., Новосибирск (2006)
10. Luethi, M.: Evaluation of energy saving strategies in heavily used rail networks by implementing an integrated real-time rescheduling system. In: WIT Transactions on the Built Environment, Computers in Railways XI, vol. 103, pp. 349–358 (2008). https://doi.org/10.2495/CR080351
11. Bocharnikov, Y.V., Tobias, A.M., Roberts, C., Hillmansen, S., Goodman, C.J.: Optimal driving strategy for traction energy saving on DC suburban railways. IET Electr. Power **1**(5), 675–682 (2007)
12. Garcia-Manriquez, J.: Modelling fuel consumption of advanced technology vehicles in traffic networks using artificial neural networks. Carleton University (Canada), AAT NQ97824 (2004)
13. Chebotarev, V., Davydov, B., Godyaev, A.: Stochastic traffic models for the adaptive train dispatching. In: Proceedings of the First International Scientific Conference Intelligent information technologies for industry (IITI'16), Advances in Intelligent Systems and Computing, pp. 323–333 (2016). https://doi.org/10.1007/978-3-319-33816-3_32
14. Davydov, B., Chebotarev, V., Kablucova, K.: Stochastic model for the real-time train rescheduling. Int. J. Transp. **1**(3), 307–317 (2017). Special Issue, COMPRAIL, Railway Engineering, Design and Operation

Decision Support System to Improve Delivery of Large and Heavy Goods by Road Transport

Anton Pashkevich[1]([⊠]), Ksenia Shubenkova[2], Irina Makarova[2],
and Damir Sabirzyanov[2]

[1] Krakow University of Technology, Kraków, Poland
apashkevich@pk.edu.pl
[2] Kazan Federal University, Naberezhnye Chelny, Russian Federation
ksenia.shubenkova@gmail.com, kamIVM@mail.ru

Abstract. The paper is devoted to special features of the large and heavy goods (oversized and overweight loads) transportation by road transport. To organize a delivery of large and heavy goods, two main problems have to be solved: selection of the optimal road vehicle as well as selection of the best route. Since there are a lot of factors have to be considered, it is necessary to use information and communication technologies. The concept of Decision Support System (DSS) together its modules' interaction is presented. Software complexes, which are based on the developed algorithm to distribute the vehicle fleet units to the suitable orders, and the set of models, which allow predicting consequences of the proposed solutions, are the intelligent heart of our DSS.

Keywords: Oversized loads · Overweight loads · Heavy goods
Large goods · Combination vehicles · Route planning · Delivery schedule

1 Introduction

Logistical costs, 30–50% of which are transportation expenses [1], are one of the main problems, which complex industries face nowadays. Therefore, as Dondo and Cerdá [2] state, such companies need to implement logistical information systems. Such systems allow to reduce the probability of errors associated with the so-called "human factor" when making such managerial decisions, as to build rational transport routes, to schedule deliveries as well as to select vehicles taking into account characteristics of the cargo. The last one is especially important, if it refers to delivery of construction cargoes by road transport. This is due to 2 reasons. Frist of all, spectrum of construction cargoes consists of several hundred different types. Secondly, each of these types requires to use its own vehicle with a specialized body, such as concrete mixing transport trucks, tipping vehicles, low-bed vehicles, flatliners, pole trailers, etc. Moreover, construction cargoes are mostly refer to the Large and Heavy Goods (LHG). LHG is a load, which exceeds the standard or ordinary legal size and/or limits of a gross combination mass or mass limits per axle and which requires special permits and handling on a designated route.

Thus, the improvement of large and heavy construction goods' transportation process relates to the problems of multi-criteria optimization, which is characterized by

G. Sierpiński (Ed.): TSTP 2018, AISC 844, pp. 13–22, 2019.
https://doi.org/10.1007/978-3-319-99477-2_2

a variety of different parameters or criteria. In this regard, traditional managerial methods are ineffective. Firstly, these tasks require the high qualification of the personnel (dispatchers and logistics specialists), who forms the criteria and data, determines the objectives and assesses the quality of recommended solutions. Secondly, they are characterized by great computational complexity [3]. Since existing approaches of directed search for the reasonable and efficient management solutions require to carry out computer-based experiments, different information systems, expert systems, decision support systems, etc. are developed.

The main goal of this article is to propose and to describe the Decision Support System (DSS), which was developed by authors. This system is directed to reduce time costs and to increase the probability of making the best decisions in difficult situations when organizing LHG's transportation. This DSS will allow to intellectualize the selection of vehicles depending on the cargo features as well as to automatize the processes of vehicle's working time calculation, scheduling, routing and filling in the transport documentation.

2 Large and Heavy Goods and Their Characteristics Influencing on Logistics Process

LHG are transported by special vehicles called heavy goods vehicles (HGV) or large goods vehicles (LGV). Their legal dimensions and weights vary depending on countries and, sometimes, on regions within one country. In European Union (EU), to such vehicles is referred any truck with a gross combination mass of over 3,500 kg, but, in the same time, its laden weight must not exceed 40 tonnes and its length – 18.75 m to cross boundaries of the EU [4]. In the UK, besides HGV and LGV, there is the term "medium goods vehicle", which is used for goods vehicles with a gross combination mass between 3.5 and 7.5 tonnes (the EU regulations classify such trucks as "large goods vehicles") [5]. According to the classification of the US Department of Transportation's Federal Highway Administration [6], standard trailers vary in length from 2.43 to 17.37 m and cannot exceed 14,969 kg; the practical gross vehicle weight restriction is determined by per-axle weight limits. In Russia, the Rules for the goods transportation by road vehicles [7] and the Instruction on transportation of oversized and heavy cargoes by motor transport on roads of the Russian Federation [8] describe the term "large goods" as cargoes, which exceed the rear point of the vehicle length by more than 2 m and/or which length in total exceeds 20 m. HGV is a vehicle, which gross combination mass exceeds 40 tonnes or per-axle load is more than 6 tonnes (when running on the roads of category B) and 10 tonnes (when running on the roads of category A).

The transportation of oversized and overweight loads includes a whole set of requirements to vehicle fleet, traffic capacity, road surface conditions, as well as cargo safety, because the organization of this process could be limited by size and load capacity of bridges, size of tunnels, presence of railway crossings, electric transmission lines and communications, and even by weather conditions and season of the year. In general, a vehicle, which exceeds the legal dimensions, must have a special permit that requires extra fees to be paid in order to use the roadways legally. The permit specifies a route, which the load must follow, as well as the dates and times, when the load can be transported. This concerns delivery of the oversized and overweight cargoes both within the country and in the case of international cargo transportation.

During LHG transportation, all the most important and significant factors must be taken into account. That means that the attention by planning processes must be focused on developing the most rational solution for cargo transportation, which allows to minimize costs as well as to perform the task as soon as possible. To plan such transportation, it is necessary to have in mind their following aspects:

- type of the vehicle's body must suit the type of cargo;
- when developing a route, it is necessary to take into account, which sections of the road network are allowed to transport goods with such characteristics;
- the cargo must be delivered on time, because the non-compliance with agreed deadlines disrupts the production plan.

Researches in the field of LHG delivery are mainly aimed to solve the Vehicle Routing Problem. Since LHG is not allowed to be transported on all roads, especially within the city, the choice of a rational and safe route requires special methods. The selection of route should be based on previous experience, type and state of the roads, cargo's dimensions and weight. Analysis of the proposed route should take into account such nuances as: level and fly-over crossings of roads, bridges, toll gates, traffic signs, electrical and other lines. It is necessary to have data about repair works on roads as well as information concerning approximate traffic loads along the route. The preferable route is the route, which does not take a lot of time and is not too long. So far as planning of routes has to consider several dozens of parameters and restrictions, there are different computer routing systems based on such methods and technologies as, for example, simulation [9], GIS with the use of linear reference system [10], expert systems [11, 12], etc.

Besides routing problems, one of the most important element in the whole LHG transportation planning process is the proper selection of vehicles in relation to the type of their bodies and to their load capacities [13]. Unfortunately, this issue has not received sufficient attention in the literature today.

3   Decision Support System in the Process of LHG Delivery by Road Transport

Bowersox and Closs [14] pointed out rightly that firms with advanced logistics systems take the following view: it is cheaper to search optimal solutions using information than to carry out non-optimal shifting of resources. Modern business starts more often to use DSS, which, in fact, is a coordinated set of data, systems, tools and technologies, software and hardware. These systems help the enterprise to collect and to process information about business and environment with the aim to justify managerial actions. For these purposes, DSSs consist usually of a wide range of methods and models, including mathematical programming, statistical analysis, theory of statistical decisions and decision-making under uncertainty, heuristic approaches, etc.

DSS in the field of LHG delivery should have the following functions:

1. Selection of vehicles taking into account characteristics of cargo.
2. Selection of vehicles taking into account the maximum permissible values of a gross combination mass and axle loads.
3. Selection of a suitable road tractor for a semi-trailer taking into account the permissible load on the fifth wheel coupling.
4. Route development taking into account technical and operational characteristics of roads and traffic loads.
5. Scheduling of a vehicle work (including the schedule of loading and unloading operations, lunch and other breaks); at the same time, it must meet the requirements to use the vehicle fleet as efficient as possible during the whole working time Tw.
6. Automated filling in the whole transport documentation.

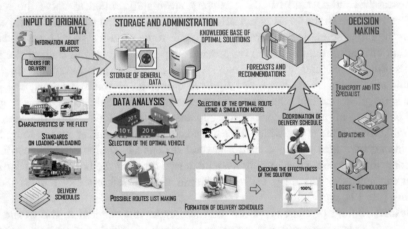

Fig. 1. Concept of the DSS's modules interaction.

In order that the system could implement all the above functions, it must have a modular structure. The scheme of interaction between Decision Support System's program modules is presented in Fig. 1.

3.1 Program Module for Data Input and Storage

This module keeps an amount of information about goods, number and structure of vehicle fleet as well as locations of all possible delivery points, which can be supplemented or changed if necessary. In addition, the database contains various database directories, the main ones of which are:

- "Tractor unit", where the information about brand and model of each road tractor is stored together with its characteristics such as: front axle load (kg), rear axle load (kg), load on the fifth wheel coupling (kg), length (m) and its state registration number.

- "Rigid truck", where the information about its brand, type of its body, load capacity (kg), length (m) and its state registration number is stored.
- "Trailer", where the information about its brand, type of its body, gross combination mass (kg), load capacity (kg), length (m) and its state registration number is stored.
- "Semi-trailer", where the information about its brand, type of its body, load on the fifth wheel coupling (kg), load capacity (kg), length (m) and its state registration number is stored.
- "Loading point", where the information about its name, address and coordinates is stored.
- "Offloading point", where the information about its name, address and coordinates is stored.
- "Cargo classification", where the information about cargo's type, overall dimensions, requirements to type of vehicle's body as well as data on other special conditions for transportation is stored.
- "Characteristics of road network", where the data on type and quality of surface, road width, permissible load, etc. is stored.

3.2   Program Module of the Optimal Vehicle Selection

The main tab in this module is the "Orders" tab, which indicates the order number, date and time of the order, date of loading, number of runs (initially this field is empty) and addresses of loading/offloading points, arrival time for loading, name and characteristics of cargo, name of customer and its phone number. When processing orders it is advisable to set the priority of long-distance offloading points, since it takes almost all vehicle's working time (Tw) to make such a transportation, and, consequently, this vehicle cannot be used for another trip on the same day.

The window concerning the optimal vehicle selection for each cargo is shown in Fig. 2. The top of the form shows the address of the offloading point, the distance in kilometers from the place of loading, weight (kg) and dimensions (m) of cargo. All procedures are performed within this window (form), the dispatcher switches between tabs, and further tabs appear when the previous actions are made.

The first tab is called "Choice of transport", where the system offers a list of all possible vehicles, which meet the criteria such as type, weight and dimensions of the cargo. In the case shown in Fig. 2, the enterprise does not have suitable free rigid trucks and trailers; however, it has two optional semi-trailers of the appropriate type and a road tractor, which can withstand the load on the fifth wheel coupling of these semi-trailers.

If there were suitable rigid trucks, trailers as well as semi-trailers, which are suitable to carry a cargo by their type of body, load capacity and dimensions, then the dispatcher would have to make a choice: (1) select a rigid truck, (2) select a combination vehicle, which consists of a rigid truck and a trailer, (3) select a combination vehicle, which consists of a road tractor and a semi-trailer. To select a road tractor with a semi-trailer, you need to select them and press the "Choose a vehicle" button, which is to the right of the tables "Tractor unit" and "Semi-trailer". To select a rigid truck, you need to select it and, without selection of a trailer, press the "Choose a vehicle" button, which is to the right of the tables "Rigid truck" and "Trailer". Therefore, to select a combination vehicle, which consists of a rigid truck and a trailer, you need to select them both.

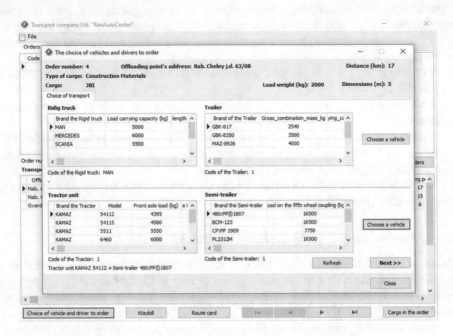

Fig. 2. Window of the program module for the optimal vehicle selection.

After the preliminary choice of the vehicle for the order, the number of runs, which need to be done, is calculated depending on the weight of the cargo and the load capacity of the vehicle. Thus, the total number of runs is automatically calculated and filled in the "Number of runs" field in the "Orders" table. The general algorithm to organize the LHG transportation is presented in Fig. 3.

3.3   Program Module of the Optimal Route Selection

So far as planning of HGV and LGV routes has to consider several dozens of parameters and restrictions, different computer routing systems based on models of region transport networks are used. However, the most software packages do not take into account the real situation on the road by routing process or take into account only current situation, which does not allow to make a long-term plan of routes. The first stage assumes formation of possible route alternatives. This could be realized by using the software package "Delivery Logistics". This package was created in mode "Managed Application" and based on the platform "1C:Enterprise" Version 8.3, which allows to distribute all orders on routes together with the minimization of total run or total delivery time. The next step is the rejection or manual correction of routes including sections of the road network, where transportation of LHG is not permitted. Then the best route can be selected from the list of remaining routes. Since the real situation on the road network including characteristics of traffic flows varies with time of day, day of week and time of year, an experiment based on the simulation model seems to be the best approach to make a choice of an optimal route alternative in urban conditions for the certain time interval.

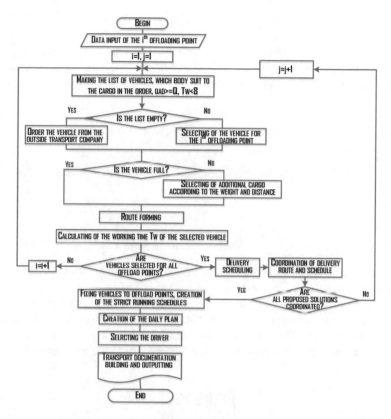

Fig. 3. Algorithm to distribute vehicles for the orders.

Program module of the optimal route selection consists of the set of simulation and optimization models. The aim of models is to minimize the total number of vehicles and the total distance of transportation. The main goal function is following:

$$Z = \sum_{k \in V_M} \sum_{(ij) \in U} C_{ij}^k \times X_{ij}^k \rightarrow \min \tag{1}$$

Where: C_{ij}^k – cost of cargo transportation from point i to point j by vehicle k; M – indicator of vehicle types (by type of body and load capacity); V_M – amount of vehicles of the same type M; U – amount of loading/offloading points; X_{ij}^k – binary variables:

$$X_{ij}^k = \begin{cases} 1, & \textit{if there is a run from } i \textit{ to } j \\ 0, & \textit{if there is a run from } j \textit{ to } i \end{cases} \tag{2}$$

The way to get the cost of cargo transportation is following:

$$C_{ij}^k = C_{tkm,M} \times q_{carr,M} \times \gamma_{carr,M} \times n_{i,M} \times V_M \tag{3}$$

Where: $C_{tkm,M}$ – cost of transportation (t-km) for vehicles of type M; $q_{carr,M}$ – load capacity for vehicles of type M; $\gamma_{carr,M}$ – index of load capacity usage for vehicles of type M; $n_{i,M}$ – total amount of runs done by vehicles of type M:

$$n_{i,M} = \sum_{k=1}^{V_M} n_{i,M}^k \tag{4}$$

$n_{i,M}^k$ is an amount of runs done by vehicle k of type M. Its calculation is based on the formula:

$$n_{i,M}^k = \left[\frac{T_{i,M}}{t_{ij} + t_{ji} + t_{load} + t_{unload}} \right]^k \tag{5}$$

Where: $T_{i,M}$ – daily working time; t_{ij} – time of forward run; t_{ji} – time of backward run; t_{load} – loading time; t_{unload} – unloading time.

The restrictions of models guarantee that (1) the daily working time will not exceeded taking into account regulations, (2) the weight of cargo will not exceeded an available load capacity, (3) minimum one vehicle will be used, (4) the index of load capacity usage will be more than 80%. All these restrictions are briefly summarized here:

$$\begin{cases} T_{i,M} \leq T_{norm} \\ W_{goods} \leq q_{carr} \\ V_M \geq 0 \\ \gamma_{carr} \geq 0.8 \end{cases} \tag{6}$$

Where: T_{norm} – permissible daily working time; W_{goods} – weight of goods, which must be transported.

The route selection algorithm and simulation model, which is the basis for decision-making, are considered in more detail in the previous paper [15].

3.4   Program Module of Optimal Solutions Storage

Since it is very difficult to describe in software all possible variants of emergency situations, when implementing the DSS, the final decision is made by a person. It could be dispatcher, logistics specialist or other decision-making person. However, the DSS should have a module (knowledge base), where these decisions done by the people on the basis of experience and intuition are recorded, preserved and, if such a situation is repeated, issued automatically as a recommended solution. It allows to choose quickly a rational variant in the case of situation repetition.

4 Conclusions

Managerial decisions in the transportation logistics must be directed to increase the sustainability of the transport system and to reduce the accidents' probability together with full satisfaction of transport needs of the population and business. Along with it, one of the areas, which requires special attention, is the LHG transportation, since its planning and organization is a labor-intensive and complex process due to the specific features of these cargoes. To ensure the safety and sustainability of the transport system when LHG transportation, it is necessary to solve two problems: to choose the vehicle, which is optimal in terms of its body type and weight characteristics, and also to develop a traffic route, which will take into account all the features of the road network, including traffic loads, and at the same time will aid to reduce logistics expenses due to optimization of supply chain management. Only complex system solutions allow to find acceptable options taking into account interests of all participants of transportation process. To make this, the DSS integrating software applications for routing and with intelligent modules, which allow to select the best type of a road vehicle and the optimal route taking into consideration the state of road network. The proposed system helps the dispatcher to make an effective decision on the distribution of the vehicle fleet for the orders and on the development of routes and delivery schedules while reducing the time and labor costs, as well as the probability of errors, which are related to the human factor.

References

1. The Establish Davis Database: Logistics Cost and Service 2015, May 2016. https://static1. squarespace.com/static/57bf65a1c534a52224df643c/t/57f6a2aa6a4963b686c1bc34/ 1475781323323/Logistics+Cost+and+Service+2015.pdf
2. Dondo, R., Cerdá, J.: The heterogeneous vehicle routing and truck scheduling problem in a multi-door cross-dock system. Comput. Chem. Eng. **76**, 42–62 (2015). https://doi.org/10. 1016/j.tre.2016.01.011
3. Oliveira, L.S., Saramago, S.F.P.: Multiobjective optimization techniques applied to engineering problems. J. Braz. Soc. Mech. Sci. Eng. (2010). http://www.scielo.br/scielo. php?script=sci_arttext&pid=S1678–58782010000100012
4. European Commission, Mobility and Transport, Road Safety, Heavy goods vehicles (2015). https://ec.europa.eu/transport/road_safety/specialist/knowledge/vehicle/safety_design_ needs/heavy_goods_vehicles_en
5. Government services NiDirect, Towing trailers with medium sized vehicles between 3.5 and 7.5 tonnes (2015). https://www.nidirect.gov.uk/articles/towing-trailers-medium-sized- vehicles-between-35-and-75-tonnes
6. The United States Department of Transportation, FHWA Vehicle Types (2015). https:// www.fhwa.dot.gov/policyinformation/

7. Rules for the goods transportation by road vehicles (in the version of the Decree of the Government of the Russian Federation of December 27, 2014, No. 1590) [In Russian: Правила перевозок грузов автомобильным транспортом (в редакции постановления Правительства Российской Федерации от 27 декабря 2014 г. N 1590)]

8. Instruction on transportation of oversized and heavy cargoes by motor transport on roads of the Russian Federation (approved by the Ministry of Transport of the Russian Federation on May 27, 1996, registered with the Ministry of Justice of the Russian Federation on August 8, 1996, No. 1146) [In Russian: Инструкция по перевозке крупногабаритных и тяжеловесных грузов автомобильным транспортом по дорогам Российской Федерации (утверждена Минтрансом РФ 27 мая 1996 г., зарегистрирована в Минюсте РФ 8 августа 1996 г. № 1146)]

9. Petru, J., Dolezel, J., Krivda, V.: Assessment of the transport routes of oversized and excessive loads in relation to the passage through roundabout. In: IOP Conference Series: Materials Science and Engineering, vol. 236, no. 1, p. 012033 (2017)

10. Dayan, S., Yoo, S., Chou, C.-S., Nichols, A.P.: GIS routing and analysis of oversize and overweight truck. In: 20th ITS World Congress, Tokyo, 109334 (2013)

11. Waheed, A., Adeli, H.: A knowledge-based system for evaluation of superload permit applications. Expert Syst. Appl. **18**(1), 51–58 (2000)

12. Jiang, G., Feng, D., Zhu, W.: A large and heavy cargo transport system designed for small and medium ship maintenance and repair. J. Ship Prod. Des. **33**(3), 212–220 (2017)

13. Ryczyński, J., Smal, T.: Proposition of a model for risk assessment in the transport of the oversized loads in the army. In: 2017 International Conference on Military Technologies (ICMT), Brno, Czech Republic, 31 May–2 June 2017 (2017). http://ieeexplore.ieee.org/stamp/stamp.jsp?arnumber=7988749&tag=1

14. Bowersox, D.J., Closs, D.J., Cooper, M.B.: Supply Chain Logistics Management. McGraw-Hill Higher Education, New York (2013)

15. Makarova, I., Khabibullin, R., Shubenkova, K., Pashkevich, A.: Logistical costs minimization for delivery of shot lots by using logistical information systems. Procedia Eng. **178**, 330–339 (2017)

Experimenting with Routes of Different Geometric Complexity in the Context of Urban Road Environment Detection from Traffic Sign Data

Zoltán Fazekas$^{(\boxtimes)}$, Gábor Balázs, László Gerencsér, and Péter Gáspár

Institute for Computer Science and Control (MTA SZTAKI), Budapest, Hungary
{zoltan.fazekas,gabor.balazs,laszlo.gerencser,
peter.gaspar}@sztaki.mta.hu

Abstract. Traffic sign data is used as input for detecting change in the type of urban road environment in which an ego-car is driven. The automatic urban road environment type detection is seen as a useful advanced driver assistance systems (ADAS) function that could be implemented in a straightforward manner relying on an existing ADAS function. Concretely, the traffic signs encountered along the route could be detected and logged by an on-board camera-based traffic sign recognition (TSR) system; and in order to perform the required function, it could be augmented with the change detection method described in the paper. Based on the analysis of the traffic sign and car trajectory data gathered for this preliminary study, the empirical distributions of the traffic signs along routes taken depend also on the geometrical complexity of these routes. A convenient model for describing traffic sign data along a route is a marked Poisson process. A traffic sign log is seen as a realization of such a process, and the minimum description length principle is harnessed to detect change in the road environment type. To test the applicability of this approach for urban routes of different complexity, a number of road environment transitions are looked at, while taking the route complexity into consideration. Finally, the detected changes are compared to ground truth.

Keywords: Urban environments · Autonomous vehicles · Change detection
Statistical inference · Poisson processes · Random walk · Fractals

1 Introduction

Car-drivers are assisted in a number of ways in perceiving and understanding the driving conditions and the actual road environment (RE) surrounding their cars. Drivers of smart cars are assisted also by the advanced driver assistance systems (ADAS) installed in their cars [1]. Among the various ADAS functions the traffic sign recognition (TSR) is of particular interest regarding the application presented herein. An overview of the TSR approaches, as well as a comparison of camera-based and other sensor-based traffic sign detection possibilities are given in [2, 3]. Moving away from the anthropocentric ADAS solutions to fully autonomous car-control mechanisms, a

© Springer Nature Switzerland AG 2019
G. Sierpiński (Ed.): TSTP 2018, AISC 844, pp. 23–32, 2019.
https://doi.org/10.1007/978-3-319-99477-2_3

versatile architecture for the evaluation of different types of ADAS subsystems – with an emphatic view on the fully autonomous driving and its related functions – was proposed in [4].

In the context of urban RE detection (RoED), the traffic sign (TS) data, which could well be gathered by a TSR system, is analyzed herein using a statistical change detection method with the intention of detecting the urban RE type, or more precisely a transition between two REs along a route, in an automatic manner. This intention is motivated by the fact that in different REs, there are different things (e.g., vehicles, obstacles) and human road-users in a variety of roles for a car-driver, be it a human being, or a robot, to look out for. For instance, a car-driver should drive more cautiously in a busy downtown area, than, say, in a calm residential district with virtually no traffic.

To support drivers in the above tasks, a RoED subsystem was proposed in [5]. The actual implementation was tested for rural and intercity environments. The subsystem makes use of the advanced perception methodology and of the up-to-date vehicle automation techniques, as well as recent inter-vehicular and vehicle-to-infrastructure communication means and methodology to achieve real-time detection and classification of obstacles and also to identify other potential risks also in real-time.

A pragmatic RoED approach was proposed in [6]. The subsystem based on this approach constitutes a low-cost surrogate for the comprehensive, computer vision-based RoED ADAS subsystems. The approach makes use of TSs that have been encountered – i.e., detected, recognized, located and logged – along a route. Routes are used for two different purposes here, namely for data collection and for actual RoED. The availability of a TSR ADAS subsystem on-board is crucial to the practical implementation of the approach. Then again, the RoED ADAS function, as well as its TSR-based implementation, can facilitate the computer vision and image understanding computations – taking place in the other ADAS subsystems – by providing geometrical constraints and regions-of-interests for these; in this manner, the availability of the RoED subsystem would make the various ADAS-related detections more robust and reliable.

Herein, the above TSR-based approach is adopted for the purpose of urban RoED, and is further analyzed in respect of routes of different geometrical complexity.

The rest of the paper is organized as follows. In Sect. 2, the relationship between the REs, the TSs installed there and the routes leading through the given area is discussed. In Sect. 3, the mathematical background of the statistical change detection approach is outlined, while the details of the TS and RE data collection activities are given in Sect. 4. In Sect. 5, the dependence of the empirical TS statistics on the route complexity is attested with the help of diagrams. In Sect. 6, the results are assessed, further work is outlined and conclusions are drawn.

2   Road Environments, Straight vs. Tangled Routes and Traffic Signs

Experience shows that certain traffic and information signs appear more frequently in urban downtown areas than they do elsewhere. These include railway/bus station signs, restaurant, hotel/motel, cafeteria/refreshments, museum/historic building signs, furthermore signs indicating parking places in the vicinity and ones warning the drivers of pedestrian traffic. Other TSs appear more frequently in industrial/commercial areas, or in residential areas. The mentioned three urban RE types were considered in [6] and these types are considered also herein. In Fig. 1, a typical scenery – at least for Hungary – is shown for each of these RE types. More common TSs, such as the ones shown in Fig. 3, are however, seem better choices for identifying the actual RE types than the ones mentioned above. This is because there is no need then to find an urban location with a cafeteria, a railway station, etc. first.

Fig. 1. Photos of a downtown area in Csepel (left), of an industrial/commercial area in Százhalombatta, (middle), and of a residential area in Vác, Hungary, respectively.

Besides the TS data, one could make use of other data and data sources readily available in smart cars for the purpose of RoED. From the many possibilities, the measurement data from the steering and speed sensors could be mentioned here as a first choice. Based on such data one could compute the car trajectory and its geometrical complexity in a real-time manner. The geometrical complexity of the car trajectory is taken into consideration and experimented with in the RoED approach presented herein.

The geometrical complexity of a trajectory can be expressed, for example, with its current fractal dimension. Examples of relatively straight, curly and tangled trails are presented in [7] and the corresponding fractal dimensions are also given there.

Having information on geometrical complexity of the route, furthermore, understanding how routes of different complexity probe and explore an urban area can be beneficial in designing and implementing a TSR-based RoED subsystem.

When driving along a relatively straight route, one is likely to use the main roads primarily and to make only a few turns. As an example, the longer, relatively straight trajectory segments of a data collection route in Csepel, Hungary are marked dark grey in Fig. 2.

The tangled routes, on the other hand, are characterized with many turns and – from time to time – even loops can occur. Driving along these routes reveal many of the nooks and corners of the urban texture to the car driver. The tangled trajectory segments of the mentioned data collection trip are presented using a lighter shade in Fig. 2.

Now, a few words about the color scheme used herein for the diagrams and figures. In Figs. 1, 3, 5 and 6, the urban environments considered in the paper, i.e., downtown (Dt), industrial/commercial (Ind) and residential (Res) areas, and any photos and values related to these environments are marked/framed with different shades of grey in a consistent manner. E.g., pairs of grey shades are used in Figs. 5 and 6 indicate the particular urban RE transitions encountered, namely Dt → Res, and Ind → Dt, respectively.

Fig. 2. The trajectory of a data collection car trip in Csepel, Hungary. The relatively straight crosstown trajectory segments are printed dark grey, while the tangled trajectory segments are presented using somewhat lighter shade. The roads, railways and waterways map layers are also shown. (Source of geographic data: OpenStreetMap, map editor: QGIS 2.8)

3   Mathematical Background

3.1   Marked Point Processes

A convenient model for describing TS data is a marked point process. A point process is customarily given by an increasing sequence of time points, say T_n. However, the TSs that appear one after the other along a route are more conveniently described in space than in time. For this reason, the driving distance, or path-length, rather than time was chosen to characterize the point process corresponding to TS data. The points of a point process may be labelled with marks. A marked point process then can be formalized as a pair (T_n, ρ_n), where ρ_n is the mark. For instance, in a log of TSs, a TS location – expressed with the path-length of the route – may carry a label stating the TS type. In many practical cases that involve marked point processes, the marked Poisson processes have proved convenient and flexible. Although the Poisson process is a continuous-time – or continuous space – model, in the algorithm presented here its discrete space approximation was applied.

3.2   Change Detection

The problem of detecting abrupt changes in the dynamics of stochastic signals has been widely discussed in the literature together with important applications. Initially, the change detection within independent and identically distributed random data was the main target of research. This effort led to the well-known Page-Hinkley change detector (PHCD), see [8–10]. The PHCD was later adopted and analyzed also for dependent data. The most important performance criteria for a change detector are its average run length between false alarms and the expected delay in detection.

3.3 Minimum Description Length Approach in Change Detection

A novel approach to change detection based on the minimum description length principle (MDL principle) – the MDL principle itself was proposed in [11] – was suggested in [12]. The MDL principle has its theoretical foundations in information theory. The stochastic approach based on this principle is used with success in various tasks, such as model selection, feature extraction, as well as in certain summarizing tasks, see [13, 14]. The basic idea of the MDL approach is to choose between models for describing data on the basis of the minimum code-length by which one can encode the data relying on these models. The advantage of the MDL methodology is its enormous flexibility; e.g., the widely used PHCD can be interpreted as a procedure relying on this approach.

3.4 The Page-Hinkley Change Detector

Assume that we have a sequence of observations ξ_1, ..., ξ_N, which is composed of two parts. The first part of the sequence is an independent identically distributed (iid) sequence of random variables taking discrete values according to the probability law p (ξ_n, θ_1), while the rest of the sequence is generated according to another probability law $p(\xi_n, \theta_2)$. The problem is then to estimate the time/location of the change between the two probability laws from observed data in a real-time manner.

The MDL approach to solve this problem is as follows. Choose an arbitrary time/location τ, and assuming that this is when – or where – the transition between the probability laws takes place, encode the observed data optimally using hypotheses concerning the data generating mechanism. Based on the standard results of information theory, the overall optimal code-length $L_N(\tau)$ of the observed data – in an asymptotic sense and allowing block coding – is given below.

$$L_N(\tau) = \sum_{n=1}^{\tau-1} -\log p(\xi_n, \theta_1) + \sum_{n=\tau}^{N} -\log p(\xi_n, \theta_2) \tag{1}$$

Following the MDL principle, the estimator of the transition time/location is obtained by minimizing the overall optimal code-length in τ. A heuristic procedure for minimizing L_N in real-time is obtained by identifying the time-point/the location after which L_N has a definite upward trend. Reformulating this characteristics of L_N, one gets a sequence which is typically 0 before the true change-point, and typically increasing after that. This is actually the output of the well-known PHCD when the detector is applied to detecting the change in the parameter of the probability law. The mentioned heuristic procedure was elaborated for the discrete time/location approximation of a marked Poisson process, as well. It was obtained by assuming an inhomogeneous iid sequence of random variables with binomial distribution, taking values 1 and 0, with probabilities θ_i and $1 - \theta_i$, respectively. Here, index i is the sequential number of the stochastic model, or in the given case, the sequential number of the urban RE, e.g., downtown area: 1, residential area: 2, and industrial/commercial area: 3. The difference of the optimal code-lengths encoding the j-th observation using the two respective probability laws is as follows:

$$\Delta L(j) = -\xi_j \cdot \log\frac{\theta_1}{\theta_2} - (1 - \xi_j) \cdot \log\frac{1 - \theta_1}{1 - \theta_2} - \sum_{k=1}^{m} \xi_j \cdot \zeta_{j,k} \cdot \log\frac{p_{1,k}}{p_{2,k}} \qquad (2)$$

Here, probabilities θ_1 and θ_2 pertain specifically to downtown areas and residential areas, respectively. Similar formulas can be derived for other pairs of REs. Relying on these formulae, one can compute the overall optimal code-length (assuming a specific pair of REs).

The aforementioned minimization is achieved using the PHCD to find the estimator of change-point τ. The specific signal to be monitored by the detector is given below.

$$g_n = S_n - \min_{m \leq n} S_n, \text{ where } S_n = \sum_{k=1}^{n} \Delta L_N(k) \qquad (3)$$

The outputs of PHCDs used for detecting change in the urban RE type based on the collected TS data and considering also the route complexity are shown in Figs. 5 and 6 as examples.

4 Data Collection from Urban Areas

In the frame of a pilot study, a car-based data collection was carried out in respect of TSs and urban REs in three urban areas in Hungary. The areas involved were Csepel (a district of Budapest), and the towns of Vác and Százhalombatta. The environments considered are Dt areas featuring one-, or multi-story buildings built next to, or very close to each other; Res areas featuring green spaces and one- and two-story buildings with somewhat more space between neighboring buildings; and Ind areas featuring factory buildings, workshops, stores with rather spacious yards, as well as shopping malls with parking lots.

The A tablet-based Android application was used for entering the TS and urban RE data manually, while the trajectory data was collected automatically by the application.

The collected TS and urban RE logs were later post-processed to separate the relatively straight and the tangled route segments. Figure 2 presents a post-processed route.

5 Detecting Change in the Urban Road Environment Based on Traffic Sign Data

Based on the collected TS and urban RE data, the four TS types shown in Fig. 3 were found suitable for separating the three urban environment types and to detect the transitions between them. The utility of these TSs rests upon their frequent occurrences in these environments, as well as on the sizeable differences between their empirical frequencies, as it is shown in Fig. 3.

The recorded logs of TSs are seen as realizations of a marked inhomogeneous Poisson point process, which was approximated in discrete space with a marked inhomogeneous binomial point process as described in Subsect. 3.4. A subset of logs was used to tune the model parameters incorporated in the change detectors, while a different subset was used for testing.

Table 1. Composition of the car routes in respect of REs and route complexity

RE type/route complexity	Dt	Ind	Res	Altogether
Relatively straight	9.9 km	30.9 km	15.4 km	56.2 km
Tangled	7.8 km	14.8 km	21.7 km	44.3 km
Altogether	17.7 km	45.7 km	37.1 km	100.5 km

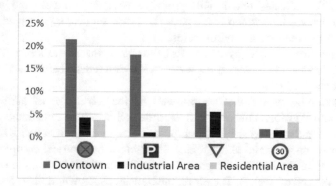

Fig. 3. Empirical probabilities of the indicated TS occurrences – along a 50 m path-length – in Dt, Ind and Res urban REs based on all the data collected.

The composition of the routes in respect of the various RE types and the geometrical complexity of the route is given in Table 1.

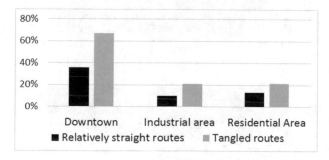

Fig. 4. Empirical probabilities of the aggregated TS occurrences – along a 50 m path-length – in Dt, Ind and Res REs, while driving along a relatively straight and a tangled route, respectively, based on all the relevant TS data collected.

In the concrete implementation of the change detection method, the car trajectory was broken up into 50 m long trajectory segments and PHCDs – one for each kind of the urban RE transitions – were employed for detecting change along the route.

The possible environmental transitions are as follows: from a Res area to Dt (Res → Dt), and vice versa (Dt → Res); from the Dt to an Ind area (Dt → Ind), and vice versa (Ind → Dt); and from an Ind area to Res area (Ind → Res), and vice versa (Res → Ind).

The PHCDs are triggered by the lack of – i.e., $(1 - \xi_j)$ in (2) – or the occurrence of a TS – i.e., ξ_j in (2) – in each segment. The segments are indexed with j. The parameters of the detectors must be tuned to the respective empirical probabilities in the manner described in [3]. These empirical probabilities are shown in Fig. 3.

Based on these empirical probabilities, one expects that the Dt areas will be easily distinguishable from the Res and Ind areas, as the TSs located in the first have a rather distinctive distribution – over the TS categories – compared to the latter two areas. On the other hand, the Ind and the Res areas have quite similar distributions over the TS categories. This indicates that it will probably be more difficult and will involve considerable delay – both in terms of space (i.e., covered driving distance) and time – to distinguish between these environments.

The geometrical complexity of the RoED route being an important factor in the change detection task is indicated by the sizeable differences between TS occurrence probabilities shown in Fig. 4.

The functioning of the PHCDs, as well as the advantage of taking the route complexity into consideration in the change detection are demonstrated in this sub-section via presenting various PHCD outputs for two particular urban RE transitions, which had been encountered along routes of different geometrical complexity.

The outputs of the PHCDs are presented diagrammatically using the color scheme described at the end of Sect. 2. Therefore, the transitions shown in Figs. 5 and 6 should be easy to grasp. The path-lengths given in the figures were calculated from the start of the given data collection trip. The TS locations along the route – for the TS types shown in Fig. 3 – are marked with small triangles along horizontal axes.

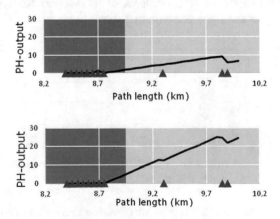

Fig. 5. Output of two differently tuned PHCDs for a particular Dt → Res transition in Csepel that was recorded along a tangled RoED route. See details in the text.

In Fig. 5, change detection results for a particular Dt → Res transition in Csepel that was recorded along a tangled RoED route is presented in the form of two PHCD output signals. The PHCD that produced the signal presented in the upper diagram of Fig. 5 was tuned to the empirical TS probabilities computed for the relatively straight route segments, and the lower diagram in corresponds to the data coming from the tangled routes. The output signal more indicative of the RE change is the second one, as it indeed should be the case.

In Fig. 6, a similar arrangement of the diagrams is presented for a particular Ind Dt transition in Csepel that was recorded along a relatively straight route. Again, the output signal of the PHCD whose tuning – in respect of route complexity and the RE transition type – is matching those the actual RoED route, respectively, is the most sensitive to the urban RE change.

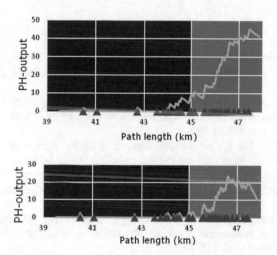

Fig. 6. Output of two PHCDs for an Ind → Dt transition in Csepel that was recorded along a relatively straight route.

The results shown in Figs. 5 and 6 rely on the knowledge of the RoED route's geometrical complexity. This information could computed from the ego-vehicle's steering and speed data in real-time as discussed in Sect. 2.

6 Conclusions

When driving from one area to another within an urban settlement the character of the RE might change considerably. The change observed along the route is often reflected also in the spatial distribution of the TSs. Taking logs of TSs from within the ego-car can be considered probing the spatial distribution of the TSs – which are installed in the area – along a particular route. These logs could be utilized to identify the actual RE type, and to locate a transition point between different REs. Herein, a circumstance of this probing is analyzed and discussed, namely whether the geometrical complexity of the route influences the RE change detection results and performance. According to the results of the small-scale study presented here, this circumstance is important, and if taken into consideration, it improves the localization of the change-points. In order to make the TSR-based RoED technology usable in the automotive industry, setting up extensive dedicated databases for storing TS and urban environment data would be essential and it is imperative that the relevant data comes from a wide range of regions and countries.

Acknowledgements. This work was supported by the National Research, Development and Innovation Fund through the project 'SEPPAC: Safety and Economic Platform for Partially Automated Commercial Vehicles' (VKSZ 14-1-2015-0125).

References

1. Bengler, K., Dietmayer, K., Farber, B., Maurer, M., Stiller, C., Winner, H.: Three decades of driver assistance systems: review and future perspectives. IEEE Intell. Transp. Syst. Mag. **6** (4), 6–22 (2014)
2. García-Garrido, M.A., Ocana, M., Llorca, D.F., Arroyo, E., Pozuelo, J., Gavilán, M.: Complete vision-based traffic sign recognition supported by an I2V communication system. Sensors **12**(2), 1148–1169 (2012)
3. Forczmański, P., Małecki, K.: Selected aspects of traffic signs recognition: visual versus RFID approach. In: International Conference on Transport Systems Telematics, pp. 268–274. Springer, Berlin (2013)
4. Belbachir, A.: An embedded testbed architecture to evaluate autonomous car driving. Intell. Serv. Robot. **10**, 109–119 (2017)
5. Jiménez, F., Naranjo, J.E., Anaya, J.J., García, F., Ponz, A., Armingol, J.M.: Advanced driver assistance system for road environments to improve safety and efficiency. Transp. Res. Procedia **14**, 2245–2254 (2016)
6. Fazekas, Z., Balázs, G., Gerencsér, L., Gáspár, P.: Inferring the actual urban road environment from traffic sign data using a minimum description length approach. Transp. Res. Procedia **27**, 516–523 (2017)
7. Katz, M.J., George, E.B.: Fractals and the analysis of growth paths. Bull. Math. Biol. **47**, 273–286 (1985)
8. Page, E.S.: Continuous inspection schemes. Biometrika **41**, 100–115 (1954)
9. Hinkley, D.: Inference about the change-point from cumulative sum tests. Biometrika **58**, 509–523 (1971)
10. Lorden, G.: Procedures for reacting to a change in distribution. Ann. Math. Stat. **42**, 1897–1908 (1971)
11. Rissanen, J.: Modeling by shortest data description. Automatica **14**, 465–658 (1978)
12. Baikovicius, J., Gerencsér, L.: Change point detection in a stochastic complexity framework. In: 29th IEEE Conference on Decision and Control, pp. 3554–3555 (1990)
13. Lakshmanan, L.V.S., Ng, R.T., Wang, C.X., Zhou, X., Johnson, T.J.: The generalized MDL approach for summarization. In: 28th International Conference on Very Large Data Bases, pp. 766–777 (2002)
14. Kiernan, J., Terzi, E.: Constructing comprehensive summaries of large event sequences. ACM Trans. Knowl. Discov. Data **3**, Article-Number 21 (2009)

Road Traffic Scene Acquisition

Ireneusz Celiński[✉], Jerzy Łukasik, Szymon Surma,
and Jakub Młyńczak

Faculty of Transport, Silesian University of Technology, Katowice, Poland
{ireneusz.celinski, jerzy.lukasik, szymon.surma,
jakub.mlynczak}@polsl.pl

Abstract. The article describes how specifically a road traffic scene is perceived by traffic participants. Every vehicle driver perceives a road traffic scene
slightly differently, even though they are all obliged to follow a set of identical
regulations. Normative (imposed by the law) behaviours of each driver are
accompanied by other behaviours which, when combined, affect quality and
safety of traffic. In both categories of behaviours, i.e. normative and other, there
are differences that can and should be studied. On account of the broad range of
problems, the article outlines selected aspects of the subject in question. The
observations provided in the article are based on a study of vehicle drivers using
an eye tracking device. Tests were performed near railway intersections. The
observations discussed in the paper are of preliminary nature, which is due to the
small size of the test sample resulting from the testing method used.

Keywords: Traffic behaviours · Eye tracking

1 Introduction

The manner in which the road network is used depends on how well its users can adapt
to a specific set of regulations. Traffic law regulates a vast majority of vehicle drivers'
behaviour patterns. These behaviours can be described as normative. These behaviours
may include: driving with belts fastened, adapting the driving parameters to road signs
and traffic lights, following the principles of interaction with other traffic participants,
especially pedestrians, etc. In a decided majority of cases, these behaviours can be
measured to some extent using modern measurement techniques. And eye tracking is
such a technique. Acquisition of a scene of motion can be measured in many planes,
and they will all be discussed further on in the article. The overall body of processes of
conscious perception of the environment by a vehicle driver comprises what is referred
to as acquisition of road traffic scene objects (Fig. 1a and b). Acquisition is more than
just perception of objects in front of a vehicle, and it encompasses nearly the entire
hemisphere before the driver's eyes. Depending on the location in the hemisphere,
individual scene objects are perceived to a different extent and with different acquisition time. The acquisition itself involves continuous adaptation of the physical
parameters of vision to the temporary conditions of environment and motion (Fig. 1c).

Besides processes connected with the road traffic scene acquisition in relation to
normative behaviours, there are also other behaviours, as they are referred to. Other

G. Sierpiński (Ed.): TSTP 2018, AISC 844, pp. 33–48, 2019.
https://doi.org/10.1007/978-3-319-99477-2_4

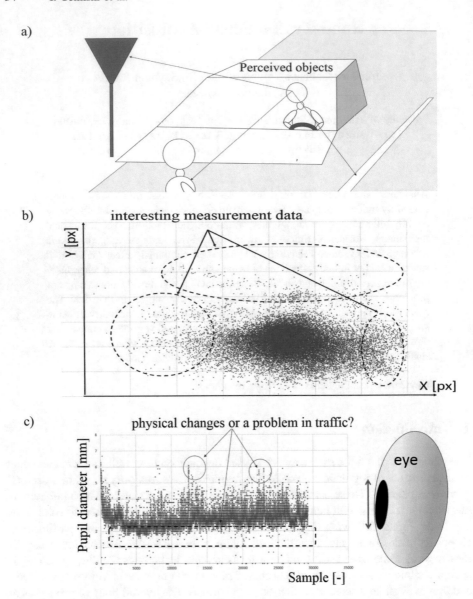

Fig. 1. Road traffic scene acquisition: (a) objects perceived, (b) plane describing fixation points, (c) physical adaptation of parameters of vision, pupil diameter. Source: authors' own research, including output from BeGaze 3.4.

behaviours are related to the vehicle driver's activities not directly connected with traffic and the vehicle use. They include telephone calls, eating while driving, speaking to the passenger etc. It should be noted that both normative and other behaviours affect traffic safety. The authors have dedicated this article to selected normative behaviours in particular. With regard to the classification of driving behaviour patterns, one more aspect requires attention. In cases of normative behaviours, all or almost all of them are related to conscious processes. Other behaviours, on the other hand, also comprise unconscious behaviours which have not yet been studied in the context of road traffic, for obvious reasons.

A road traffic scene is never a single frame describing the vehicle's environment, as it is perceived by the driver. In road traffic, such a scene comprises all frames that form a set of pieces of information that are necessary for appropriate perception of a selected infrastructure element (inlet, junction, intersection). For a road junction, a set of these frames should encompass all signs relevant for correct passage of a vehicle. What the road traffic scene acquisition requires is, among other elements, performing normative behaviours enabling a minimum information set to be built:

$$ASRD = \{ZDP, ZDO, SD, PUR, IURK, GI, ID\} \tag{1}$$

where:

ZDP – set of vertical traffic signs in the scene
ZDO – set of horizontal traffic signs in the scene
SD – set of traffic signals
PUR – set of pedestrians participating in traffic, important from the driver's perspective
IURK – set of vehicular traffic participants
GI – set of elements of the infrastructure geometry, e.g. number of inlets etc.
ID – set of other data related to the current traffic situation.

The set described above is updated in a certain interval and in a defined hemisphere space surrounding the vehicle driver. At the same time, every element of a set noted as Eq. (1) is collected at the given instant in different sections of the sphere, with different times of the process initiation by different vehicle drivers. Each element added to the set is characterised by its type, location and recording time.

Where the foregoing is the case, road traffic scene acquisition pertains to hundreds, if not thousands of elements collected in short periods of time. As a result of the aforementioned characteristics, the eye tracking technology is currently the best solution that enables gathering information on the ways in which road traffic scene acquisition is performed. In contemporary mobile tests, perception of scene elements are recorded at the frequency of 120 Hz in binocular mode and HD resolution for traffic scene camera. The following images, graphs and table was presented in an article summarizing a study in mobile tests with the participation of more then ten peoples and vehicles. Due to privacy protection presented photos concern test drives of one of the co-authors.

2 Process of Scene Perception by a Vehicle Driver

The author of publication [1] believes that there is no pure vision, which could be referred to as "proportional to a larger area of the scene" (non-selective vision). The sense of sight always provides the user with only partial representations (objects, fragments) of the scene of motion being observed. Due to the specificity of the retina, the maximum of 5° of the total scene is clearly recorded (in B5 page format this area is approximately equal to 1 cm²). The driver's sight is even more selective on account of the vehicle running speed (Fig. 2a, b). In Fig. 2b, it is not the moon above the road, but rather the temporary centre of the driver's sharp vision area that has been marked. Only selected pieces of information directly relative to the front of perceiving person are represented in a scene subject to acquisition. In the dark there is a reversal of the process, it is easier to capture changes in the periphery. Not until a set of selective representations is processed in a certain time does it create a road traffic scene (while the vehicle is moving all the time). In road traffic, this kind of perception is affected by speed limits which constrain the choice of available information as the driving speed increases (restricting the cone of sharp vision).

Fig. 2. Examples of restrictions to road traffic scene perception: (a, b) restriction of the field of sharp view (a-filtered, b-original), (c, d) selective choice of objects in a scene. Source: author's output from the SMI BeGaze 3.4.

The choice of information (items) relevant to what is referred to as the driver's affordance basically depends on a number of factors such as the object of interest (sign, vehicle, obstacle), long- and short-term targets (choice of travelling direction, town sign interpretation etc.), degree of complexity of a stimulus (number of vehicles and/or pedestrians relevant for the path of further driving route) etc. Within one minute of driving on the road, up to 50 impacts occur. Objects outside the centre of attention may be represented, but they are not visually experienced. The process of vision is dynamic, especially in road traffic, and therefore it is not contemplative but exploratory in nature

[1]. This determines individual person's limits of perception of a road traffic scene. It is possible that both location and design of road signs does not take all such restrictions into account.

The dynamics of the vision process (change of fixation between objects) at high running speeds is often below the average response time of the population (0.25–0.90 s). For objects that either dynamically change their position in the road traffic scene or are poorly visible, the mind also formulates hypotheses on the scene being perceived and consistently directs eyeballs in such a manner as to verify the hypothesis [1, 6]. Therefore, taking action while in motion is not something that follows an observation of a road traffic scene. In road traffic, observation is an integral part of the actions performed, where objects of the road traffic scene change dynamically, often in time shorter than time of response. Figure 2c and d illustrate how dynamically and selectively sight is moved between different important elements of the traffic scene. Moreover, the road traffic scene acquisition depends on the vehicle driver's previous experience as well as knowledge (for example obligatory trips). One cannot tell any difference between the stage of the road traffic scene perception and its understanding, since both form an integral and dynamic process [1] (Fig. 3).

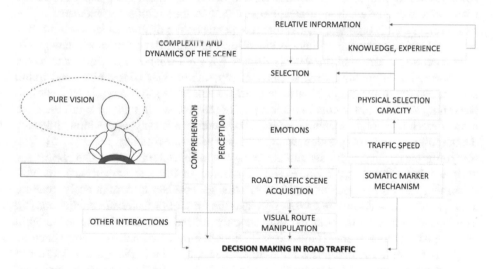

Fig. 3. Restrictions and variable to road traffic scene acquisition.   Source: author's own materials.

Also emotions affect the process of selective road traffic scene perception, as described above [2]. This translates into processes of road traffic scene analysis and consequential decision making. Emotions such as anger or aggression increase the certainty of judgements, which leads to what is referred to as shallow thinking [1]. Persons displaying such emotional states are driven by various heuristics, which consequently makes them jump to conclusions more quickly, yet without detailed analysis of all premises of the decision being made [1]. On the other hand, an emotion such as fear or uncertainty makes you inclined towards analytical thinking, although on the other hand, it delays decisions and can slow down reaction to certain elements of the road traffic scene which require rapid preventive measures. Emotions can be studied to a limited extent using the ET technique. A popular psychological hypothesis is referred to as the somatic marker mechanism. Markers develop as one acquires personal experience through social interactions. Consequently, somatic markers are acquired by learning instead of being inborn. The marker mechanism allows one to make unconscious intuitive decisions, which may be rational. A normally functioning brain activates certain states in a situation where decisions must be made [1, 3]. This leads to categorisation of alternative ways of behaving in a road scene. This information is used by the organism and allows for optimum measures to be undertaken in a dynamic environment. There is a number of behaviours that people display in road traffic that have been learned not only in driving lessons, but also in everyday use of the road network. Many of them have not been properly regulated in normative terms. The aforementioned limitations to the road traffic scene perception constitute a specific filter for potential acquisition options. However, in light of the theory of evolution, one may assume that representation of a motion scene does not need to accurately reflect its structure in order to be useful for the vehicle driver [1]. From the perspective of the road traffic scene acquisition, it means that a certain minimum set of elements of the road traffic scene is represented, and that it depends on the driver to a large extent. It is also a set which meets the usefulness condition from the driver's point of view [1, 4]. This condition means that the vehicle drivers have information allowing them to act effectively, i.e. to continue driving (most likely safely) over consecutive road traffic scenes. Motion scene acquisition is not tantamount to passive reception of stimuli. Becoming aware of traffic conditions is a part of the process. In trails of sight, there are more descending than ascending paths, and there is not only acquisition, but also decomposition and composition of motion scene images taking place [1, 5]. The foregoing means that more signals are outgoing than incoming, and that the image collected from the retina is not exactly the same as the one the driver perceives. Assuming such a perspective, the brain's strategy consists in continuously creating anticipatory road traffic scene models that take the form of probabilistic predictions [1]. The input signals acquired from eyeballs are only utilised as a part of data used to build an imaginary motion scene. If the driver is travelling in a road network with which he or she is familiar, this minimises input adjustments, and the traffic scene is known to them in a certain way. In this area there is a lot to do in the field of reconstruction of road accidents.

With reference to the discussion on the processes that affect the road traffic scene acquisition, it should be noted that different vehicle drivers may perceive the same traffic scenes differently. This is due to the fact that this process is always performed from different points in the scene, and that cognition (scene acquisition) is determined by the driver's previous experience. The knowledge acquired by individual drivers is similar in the normative sense (only the minimum exists), while it is completely different in the sphere of experience, and this is not only on account of the driver's experience, but also due to the changing normative conditions, as they exert secondary impact on the experience. Even in one population, there are different driving styles which determine different experiences (traffic in Warsaw vs Katowice). The driver's mind is a tool that functions similarly to hands and legs. The mind creates useful, yet frequently speculative representations of the reality in order to enable undertaking effective actions [9] (Fig. 4).

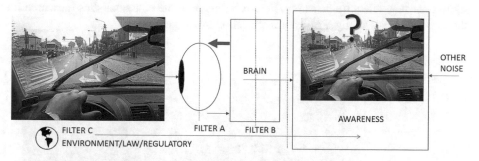

Fig. 4. Road traffic scene – image acquisition using filters: eye's and brain's physical filter and sociological and others filter. Source: author's own materials.

What also matters for the road traffic scene acquisition is social cognition, as it is referred to, which is related to the mechanism of attributing certain intentions and beliefs to other road users. In contrast to other social interactions, this process runs very dynamically. Some authors elaborating on this body of problems claim that people, and so drivers a swell, are imitation machines [1, 7–10]. One of the basic human learning strategies is indeed imitation (while in motion, e.g. giving way to others, changing lane). The ability to read other people's intentions made it possible for the cause and effect relationship category to be developed with regard to road traffic situations. Where this is the case, it causes that both bad and good driving patterns are duplicated in road traffic.

3 Examples of Measurements of Scene Acquisition Parameters

With regard to the foregoing observations, it may be claimed that the road traffic scene analysis should mainly consider relevant representations of the road traffic scene. These representations result from normative traffic regulations. Observation should take into consideration both the speed of the moving object and the associated limitations of the cone of sharp vision. The testing method should comprise the dynamics of acquisition of elements of the road traffic scene representation. The analyses being conducted under the research should enable assessment of changes in the perception of the motion scene over time.

The first measuring procedure performed by means of the eye tracking technique consists in measuring the time span between the moment when the given road sign was originally noticed and the moment when it had appeared within the scene. The road sign appears in the driver's field of view at a certain instant (assumed arbitrarily), but not until a certain time interval lapses is it actually noticed by the driver. This time is defined as the time difference from the instant when the sign appeared within the scene until the first fixation of the driver's eyes on it. What also matters for this measuring procedure is to measure time intervals until successive traffic signs are recognised (acquisition of sequence of signs). This applies to sequences of signs that can be both related and unrelated to one another. According to the authors, certain character configurations may affect traffic safety. Figure 5 illustrates the characteristics measured for the A-9 sign designating a railway intersections with gate arms.

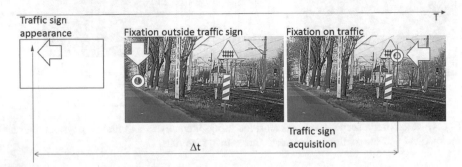

Fig. 5. Traffic sign acquisition measured by the time of fixing sight on the sign. Source: author's own materials.

Table 1 provides a collation of the acquisition time for a selected traffic sign measured using the eye tracking technique. Acquisition times were studied for four different traffic signs while heading towards a railway intersection. The test sample comprised six different drivers. The table below summarises statistical parameters established for this sample.

Table 1. Acquisition time for the chosen traffic sign [s].

Parameters	Sign 1	Sign 2	Sign 3	Sign 4
Mean t[s]	8.32	8.12	13.42	15.12
Deviation [s]	2.61	3.40	3.89	4.31

The values provided in Table 1 are derived from a video frame analysis (manual) performed in the SMI BeGaze 3.4 software. The data illustrate the time which lapses between the moment the sign appears and the moment it is noticed (fixation), also showing how variable the sample is. What proves important in this respect (with a relatively simple measuring technique) is the very high variability observed in the sample. Being able to run tests on a larger sample requires specific steps aimed at automation of the measurement method. The second procedure is measuring the amplitude of saccadic movements. This measurement illustrates how a traffic scene can be utilised from the perspective of the capacity to perceive traffic signs. With a poorly utilised road traffic scene, saccadic movements display excessive amplitude resulting from the driver's sight fixation on distracting elements of the road traffic scene. In an optimum situation, traffic signs should be positioned in their respective locations in such a manner that they do not impose excessive saccadic movements towards

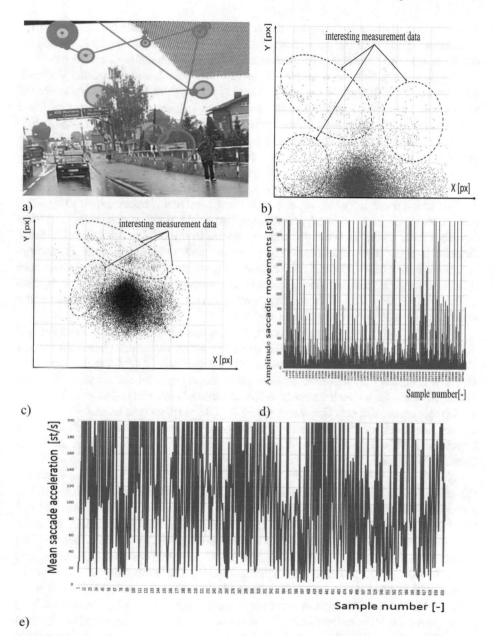

Fig. 6. Road traffic scene – sacadic movements: (a) sample scene (fixation points intentionally raised), (b) start points of saccadic movements, (c) end points of saccadic movements, (d) saccade amplitude, (e) mean saccade acceleration. Source: author's output from the SMI BeGaze 3.4.

elements potentially distracting the driver's attention. High amplitude causes driver distraction, eye fatigue and may cause skipping of relevant secondary information. In the other hand -the variability of these movements is the measure of awareness.

Figure 6 illustrates an example of important vertical and horizontal saccadic movements. What matters for the driver is to read information in a much smaller area of the motion scene. Also signs of traffic priority, pedestrian crossings, directions and town names are also important for the driver. Nevertheless, what can be clearly seen in this picture (Fig. 6a) is that large horizontal sight shifts are made – larger than the scene image would suggest. Figures 6b and c show some considerable dispersion of characteristic points associated with saccadic movements (circle).

In each of the cases shown in Fig. 7, saccadic movements significantly exceed the field in which there are traffic signs or other traffic-related elements, such as vehicles and pedestrians. The mass of distributions from Fig. 7a–d concentrates at points directly related to traffic. What one can also notice in these figures is some significant differences in terms of shifting one's sight over the road traffic scene, which the authors believe to be not substantiated by the content of this scene. What Fig. 7b reveals about the driver is a significant left-side spread of saccadic movements, and similar is the case of the driver in Fig. 7d. An interesting (even strange) separation of vertical saccadic points can be noticed in the driver in Fig. 7c.

Figure 7c and d depicts fixation points for the same road traffic scene and for two different drivers who have shown perceived the scene differently. The two picture clearly shows how different the perception of individual motion scenes which consist of identical elements may be (there was no traffic incoming from the opposite direction during the test). Every such image is in fact the driver's imprint linked with the given road traffic scene. It seems that the driver in Fig. 7d predicts how events will develop by keeping track of the road within the maximum range of sight (this but is a hypothesis). This may be due to the style of driving-frequent looking at the board (instruments panel).

Not all road traffic scene elements are consciously realised by the driver. Watching the scene, the driver picks those elements that he considers important. Some of the scene elements may be omitted when it is swept over with saccadic movements. The mind does not record all the data that have reached the brain via the eyes, unless the vehicle driver is intentionally looking for an object that may potentially become lost (heuristics problem). For a new travelling route (for example optional trip) over a road network, it is highly probable that important road traffic information can be lost, as drivers tend to view it in a rather chaotic or highly selective manner. Solutions to all these problems may be sought in analyses of road traffic scenes using the eye tracking technique which leads to developing what is referred to as heat maps and/or attention maps (Fig. 1a and b). These graphics can demonstrate what may stay potentially hidden in a road traffic scene and what may exert impact on the scene acquisition and safety. It particularly allows for checking whether, for example, the driver has laid eyes on all traffic signs, if there any marking details which the mind does not perceive, e.g. because they are outside of the path of sight, in the field of peripheral vision. To a limited extent, this case is illustrated in Fig. 8. The picture shows a prohibition sign that has always remained outside of the vehicle driver's field of sharp vision (thus being irrelevant to the latter).

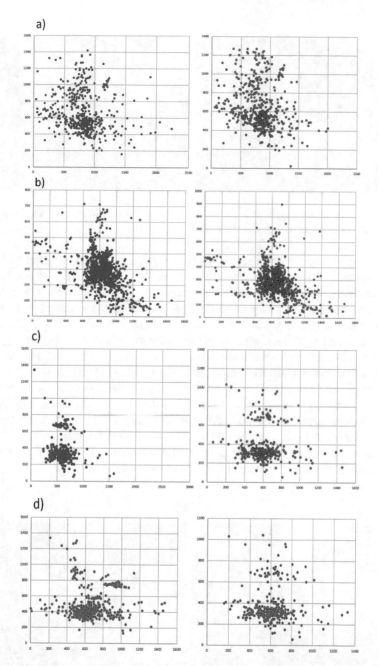

Fig. 7. Road traffic scene – sacadic movements; starting points in the left side, end points in the right side (dimension on both axes is pixels); a÷d – numbers of drivers recorded while crossing a railway intersection (X-axis in pixels, Y-axis in pixels). Source: author's output from the SMI BeGaze 3.4 program.

a)   b)

Fig. 8. Road traffic scene: (a) attention maps vs hidden content, (b) only peripheral prohibition sign noticed. Source: engineer's dissertation.

What may also prove very important in road traffic is negative inclination [11, 12], as it is commonly referred to, which is about enabling the vehicle driver to recognise dangerous elements in the road traffic scene as soon as possible. Evolutionary development promoted reactions aligned with negative inclination – a reaction to negative stimuli connected with a lethal threat. Results of tests conducted by application of video technology imply that a negative symbol accompanied by a subliminal symbol of similar content speed up the reaction. According to literature data, negative inclination is predominant throughout the entire population [11–13]. Therefore, one should investigate if there are sequences of content (traffic signs and other traffic scene elements) in the road traffic scene which may affect the speed of the driver's reaction. For example: does an advertisement promoting positive and relaxing content not reduce the response time in a vehicle driver? Further tests should be conducted to establish whether or not, and if so, in what scope and how negative content combined with other content carriers can affect the perception of a road traffic scene.

a)   b)

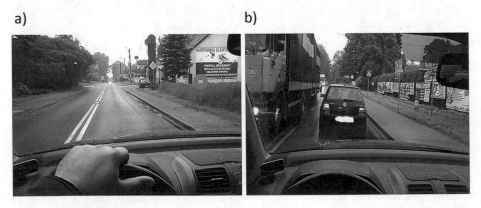

Fig. 9. Road traffic scene vs negative/positive inclination: (a) advertisement read under conditions of no traffic congestion, (b) series of advertisements (different informations) next to a vehicle moving in a congested road. Source: author's own materials.

Figure 9 shows a series of more than ten advertisements which may trigger extreme emotions in a driver (positive/negative). A specific combination of these emotions can locally reduce the response speed.

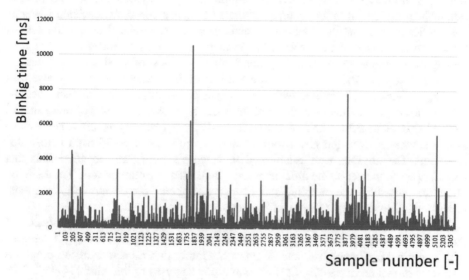

Fig. 10. Road traffic scene – blinking parameters. Source: author's output from the SMI BeGaze 3.4 program.

Besides cognitive processes, also the vehicle driver's physical reactions affect driving. Such reactions (unconditional and subconscious) may have additive or multiplicative effect on parameters of the vehicle in motion. The time of a single blink typically does not exceed 200–300 ms across the population. However, in terms of parameters of this reflex, considerable variability is recorded, especially during long journeys. Such characteristics are depicted in Fig. 10, which was prepared with reference to a record of a journey by car which took 2 h, 40 min and 10 s. During this test run, the following numerical characteristics of the data set were recorded for the driver's blinking time: 32 [ms] (min.), 10,515 [ms] (max.), 262.30 [ms] (mean), 412.10 [ms] (standard deviation). If a blink is to be treated literally as an act of closing eyes, then the conclusion is that in a 3-hour drive, the driver is steering the car with his or her eyes shut for 23.56 min in total. It should also be added that it is actually while car driving that cases of the longest continuous closing of eyes are observed, as they may reach even up to several seconds (1 - 11 [s]). Driving at average running speeds, a driver closing the eyes for 3 s loses the capacity to monitor the road over at least 30–40 metres, while at 100 km/h – the non-monitored section of the road increases to more than 83 metres. It exceeds the braking distance.

4 Conclusion

With regard to the field of research addressed in the article, one may find an interesting source of information in an experiment conducted by Benjamin Libet, who tried to establish if the human brain becomes activated before a conscious decision is made. His studies proved that the brain shows activeness in areas related to specific actions (based on a number of numerous other tests performed by application of the neuroimaging technique) even before any identifiable awareness of the decision in question emerges. Also John-Dylan Haynes conducted interesting studies using highly advanced imaging technologies supported with computer-aided measurement of human reactions (using a tomograph), thus partially confirming Libert's results. Results of these studies showed that more than a half of respondents displayed activity in brain areas commonly associated with the given form of activity before one could even realise this condition. The advance time established in Haynes's tests came to 10 s. This fact should exert an impact on the training process dedicated to future drivers. The attention of those providing the training should perhaps focus on developing situational heuristics in future drivers instead of maneuvers?

The article focuses mainly on interesting problems pertaining to the processes of the road traffic scene acquisition. Neuroscience is becoming increasingly integrated with many fields of expertise, and all premises imply that transport related problems also form in fact an interesting field for numerous analyses of this kind [14–17]. With this observation in mind, the discussion provided in this article should be brought to even more detailed level by thoroughly studying differences in personal traits that affect perception of a road traffic scene.

Observation of the driver's behavior in road traffic is not only a simple straight ahead driving test, but also complicated issues related to traffic on roundabouts [18, 19, 23]. The state of the road pavement also has an impact on the examined parameters of the driver eyes, which will be examined in the coming years more and more precisely due to the development of the ET technique [20, 21, 24]. It is also important to study drivers' behaviors in the aspect of methods of route choice and means of transport used within them [22, 25]. Interesting problems in the aspect of the presented research are provided by modern vertical parking systems [26]. In these cases, testing the driver using the ET technique requires a lot of analytical effort. The presented approach is applicable in a wide range of different traffic processes [27, 28].

It should also be remembered that the behavior of drivers change with increasing fatigue. This is a factor that has a huge impact on road scene acquisition. Fatigue testing can be performed using a variety of techniques including vision with the appropriate video sensors [29].

References

1. Brożek, B.: Myślenie, podręcznik użytkowania. Copernicus Centre Press, Kraków (2016)
2. Lerner, J.S., Li, Y., Valdesolo, P., Kassam, K.: Emotion and decision making. Ann. Rev. Psychol. **66**, 790–823 (2015)

3. Damasio, A.R., Tranel, D., Damasio, H.: Somatic markers and the Guidance of behavior: theory and preliminary testing. In: Levin, H.S., Eisenberg, H.M., Benton, A.L. (eds.) Frontal Lobe Function and Disfunction (1991)
4. Noe, A.: Action in Perception. MIT Press, Cambridge (2004)
5. Livingstone, M., Hubel, D.: Segregation of form, color, movement, and depth: anatomy, physiology, and perception. Science **240**, 740–749 (1988)
6. Dennett, D.C.: From Bacteria to Bach and Back: The Evolution of Minds. Copernicus Centre Press, Kraków (2017)
7. Tomasello, M.: Kulturowe źródła ludzkiego poznania. PIW, Warszawa (2002)
8. Brożek, B.: Rule Following. Copernicus Centre Press, Kraków (2012)
9. Chyurchland, P.S., Ramachandran, V.S., Sejnowski, T.J.: A critique of pure vision (1994)
10. Frith, Ch.: Making up the Mind. Blackwell Publisher, Oxford (2007)
11. Gazzaniga, M.S.: Human: The Science of What Makes Us Unique, New York (2009)
12. Bargh, J.A., Chaiken, S., Raymond, P., Hymes, C.: The automatic evaluation effect. J. Exp. Psychol. **32**, 104–128 (1996)
13. Rozin, P., Royzman, E.B.: Negativity bias, negativity dominance and contagion. Pers. Soc. Psychol. Rev. **5**, 296–320 (2001)
14. Libet, B., Freeman, A., Sutherland, J.K.B.: The Volitional Brain: Towards a Neuroscience of Free Will. Imprint Academic, Exeter (1999)
15. Libet, B.: Mind Time: The Temporal Factor in Consciousness. Perspectives in Cognitive Neuroscience. Harvard University Press, Cambridge (2004)
16. Dennett, D.C.: Freedom Evolves. Allen Lane, London (2003)
17. Soon, C.S., Brass, M., Heinze, H.J., Haynes, J.D.: Unconscious determinants of free decisions in the human brain. Nat. Neurosci. **11**, 543–545 (2008)
18. Macioszek, E.: The comparison of models for critical headways estimation at roundabouts. In: Macioszek, E., Sierpiński, G. (eds.) Contemporary Challenges of Transport Systems and Traffic Engineering. LNNS, vol. 2, pp. 205–219. Springer, Switzerland (2017)
19. Szczuraszek, T., Macioszek, E.: Proportion of vehicles moving freely depending on traffic volume and proportion of trucks and buses. Balt. J. Road Bridge Eng. **8**(2), 133–141 (2013)
20. Staniek, M.: Stereo vision method application to road inspection. Balt. J. Road Bridge Eng. **12**(1), 38–47 (2017)
21. Staniek, M.: Road pavement condition as a determinant of travelling comfort. In: Sierpiński, G. (ed.) Intelligent Transport Systems and Travel Behaviour. Advances in Intelligent Systems and Computing, pp. 99–107. (2017)
22. Sierpiński, G.: Technologically advanced and responsible travel planning assisted by GT Planner. In: Macioszek, E., Sierpiński, G. (eds.) Contemporary Challenges of Transport Systems and Traffic Engineering. Lecture Notes in Network and Systems, vol. 2, pp. 65–77. Springer, Heidelberg (2017). 13th Scientific and Technical Conference "Transport Systems Theory and Practice 2016", Katowice, Poland, September 19–21 2016
23. Macioszek, E., Sierpiński, G., Czapkowski, L.: Methods of modeling the bicycle traffic flows on the roundabouts. In: Mikulski, J. (ed.) Transport Systems Telematics. CCIS, vol. 104, pp. 115–124. Springer, Heidelberg (2010)
24. Staniek, M.: Moulding of travelling behaviour patterns entailing the condition of road infrastructure. In: Macioszek, E., Sierpiński, G. (eds.) Contemporary Challenges of Transport Systems and Traffic Engineering. Lecture Notes in Networks and Systems, pp. 181–191. Springer, New York (2017)

25. Sierpiński, G.: Distance and frequency of travels made with selected means of transport – a case study for the Upper Silesian conurbation (Poland). In: Sierpiński, G. (ed.) Intelligent Transport Systems and Travel Behaviour. Advances in Intelligent Systems and Computing, vol. 505, pp. 75–85. Springer, Heidelberg (2017). 13th Scientific and Technical Conference "Transport Systems Theory and Practice 2016", Katowice, Poland, September 19–21 2016
26. Pypno, C., Sierpiński, G.: Automated large capacity multi-story garage—concept and modeling of client service processes. Autom. Constr. **81**, 422–433 (2017)
27. Młyńczak, J., Celiński, I., Burdzik, R.: Effect of vibrations on the behaviour of a vehicle driver. Vibroeng. Procedia **6**, 243–247 (2015)
28. Młyńczak, J., Burdzik, R., Celiński, I.: Research on vibrations in the train driver's cabin during maneuvers operations. In: 13th Conference on Dynamical Systems Theory and Applications. DSTA (2015)
29. Małecki, K., Nowosielski, A., Forczmański, P.: Multispectral data acquisition in the assessment of driver's fatigue. In: Mikulski, J. (eds.) Smart Solutions in Today's Transport. TST 2017. Communications in Computer and Information Science, vol. 715, pp. 320-332. Springer, Cham (2017)

Junction Traffic Prediction, Using Adjacent Junction Traffic Data, Based on Neural Networks

Teresa Pamuła[✉] and Wiesław Pamuła

Faculty of Transport, Silesian University of Technology, Katowice, Poland
{teresa.pamula,wieslaw.pamula}@polsl.pl

Abstract. The paper discusses the problem of substitution of traffic data, used for prediction of traffic flows at a junction, with data from an adjacent junction. Such a case arises when the measuring resources at the junction malfunction. Neural networks based approach is used for forecasting traffic flows. Solutions incorporating a multilayer perceptron (MLP) network, a cascade forward network (CFN) and a deep learning network (DLN) with autoencoders are used for evaluating the prediction performance. The elaborated designs are validated using a data set of traffic flow measurements comprising over 15 thousand measurements collected in a period of over six months. Results prove that substituting data from an adjacent junction is justified for predicting traffic flows in case of malfunctioning measuring resources.

Keywords: Neural network structure · Traffic flow prediction
Intelligent transport system

1 Introduction

Short term prediction of road traffic parameters is crucial for traffic control and management. The precise knowledge of coming changes in the traffic scene enables a more efficient utilisation of available transport infrastructure resources [8]. This can also contribute to devise suitable measures for sustaining the safety of travel. Traffic prediction is based mostly on historical data collected at characteristic sites of the road networks. Comprehensive historical data sets combined with descriptions of traffic conditions and current data, enable predictions in time horizons from a few minutes to even few hours. These diverse data sets comprising not only numerical data are especially suitable for processing by means of artificial intelligence methods [5, 9].

Neural networks (NN) are chosen for their capability to map highly diverse behaviour of variables [4]. This mapping is recovered in the course of training the network without prior elaborate definition of the forms of relationships between the variables. This feature of NN is of particular importance as the range of behaviours is very wide and individual approaches to every case of traffic changes could be very laborious. NN have the ability to learn and generalise from data, this makes them a useful tool for working with noisy or missing data which often happens in traffic forecasting solutions.

© Springer Nature Switzerland AG 2019
G. Sierpiński (Ed.): TSTP 2018, AISC 844, pp. 49–57, 2019.
https://doi.org/10.1007/978-3-319-99477-2_5

The paper reports research results and assesses the prediction performance of three commonly used NN architectures. The networks are trained and their structural design is optimised using traffic data obtained from a Traffic Control Centre. NNs provide pairs of forecasts for adjacent junctions using data from the first in sequence junction. This is a case when vehicle detectors of the second junction malfunction or weather conditions prevent correct detections of vehicles. Proper traffic data is substituted with data from vehicle detectors operating down on the traffic route.

Section 2 of the paper summarizes related works emphasizing a need for solutions dealing with data loss. The third section presents the prediction models incorporating NNs. The next section evaluates the prediction performance. Conclusions are drawn and further research ideas are proposed in the ending section.

2   Related Works

Recent studies on applying different NN architectures with different input parameters and configurations demonstrate that NN modelling is an effective approach for short term traffic flow modelling. Multilayer perceptron (MLP), deep learning networks with autoencoders (DLN), cascade forward neural networks (CFN) and radial basis function (RBF) neural networks are the most widely used models in short-term traffic flow prediction [7, 10].

Kumars et al. [11] predict flow values, using MLP, on the basis of a sequence of historic flow values, traffic speeds and the week day for which the prediction is done. In order to achieve smaller errors the traffic stream is treated as a set of vehicles of different categories. Each category is represented by a separate input variable. Good prediction results are reported for time horizons of 5–15 min.

A similar approach is presented in [1]. Time series of traffic flow values in the morning attributed to work days, holidays are used for prediction of traffic flows in afternoon hours respectively on workdays and holidays.

The image of the traffic situation in the case of ITS is mapped with good accuracy by O-D matrices. Many works report advances in determining O-D matrices using incomplete or "noisy" traffic data from different measuring devices. Lorentzo et al. [12] developed a CFN based solution for determining the O-D matrix using traffic flow values registered between junctions. Traffic data dimensionality is reduced with PCA and the results are used as inputs of a multilayered neural network. The trained NN generates values of O-D matrices for almost real time traffic management. In [13] a solution is proposed, which uses outputs from congestion detectors, vehicle positions and time of travel for short term prediction of travel speed. The travel speed is derived from an O-D matrix. Additionally the network uses historical data in the course of training.

The accumulation and dissipation of one junction affect the state of traffic flow of adjacent junctions. These spatio-temporal features of the traffic flow changes at adjacent junction are exploited to optimize the performance of the forecasting model. In [6] authors propose, a RBF-based method by using the traffic flow data of adjacent junctions to obtain more accurate forecasting result and to solve the problem of missing data.

As yet only a few applications of DLN in traffic prediction can be noted in literature. In [14] a DLN consisting of stacked autoencoders is proposed to learn generic traffic flow features. The model successfully discovers the latent traffic flow features such as spatial and temporal correlations. A logistic regression layer on top of the network is used for supervised traffic flow prediction. Stacked autoencoders are also examined in [15] and prove to be highly effective for traffic flow prediction along a motorway.

3   Model of the Prediction Task

The control of traffic flow in road networks performed by integrated systems (ITS) relies on reliable information on road traffic parameters. The traffic data is acquired using numerous measuring devices placed along traffic routes in the network. Problems arise when some of the devices malfunction. This leads to lack of data or corrupt data, thus to distortion of the traffic image in the network.

The problem of data loss in the case of two consecutive measuring sites placed along a traffic route is investigated. Figure 1 shows the draft of this situation. It is assumed that the vehicle detectors d_1 and d_2 are placed at the approach lanes of the junctions. This is a common solution as such detectors also function as stop detectors for securing the work of local traffic controllers. The aim of the investigation is to validate the design of a traffic flow prediction module, constructed on the basis of a neural network, which provides forecasts in the situation when d_2 the second in sequence detector malfunctions.

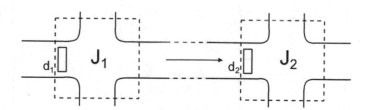

Fig. 1. Placement of detectors

The horizon of the prediction is determined by the update rate of the control parameters used by the traffic control systems. Although local control systems mostly adapt their functioning to current local traffic conditions this capability is overridden by supervisory actions when such systems are incorporated as parts of control networks (ITS). The set of update periods comprises the values 5, 15, 30 and 60 min. The most often used value is 15 min. This value amounts to ten average traffic control cycles on junctions.

The predicted traffic flow values at consecutive junctions J_1 and J_2 are a function of current and past flow values at the first junction J_1 and are worked out with the use of neural networks. The neural networks are constructed of a defined number of layers L_i, each with a set number n_i of neurons:

$$NN = L_i L_{i-1} \ldots L_1(n_1)$$
$$n_i \in (a_i, b_i), \quad i = 1, \ldots, m \tag{1}$$

where:

L_i – layer consisting of n_i neurons, the number of neurons falls in the range (a_i, b_i)
The following models of the prediction task – Fig. 2 are proposed:

– using two networks NN_1, NN_2:

$$Q_{t+1}(J_1) = f_1(Q_t(J_1), \ldots, Q_{t-n}(J_1))$$
$$Q_{t+1}(J_2) = f_2(Q_t(J_1), \ldots, Q_{t-n}(J_1)) \tag{2}$$

– using a single NN:

$$Q_{t+1}(J_1, J_2) = f(Q_t(J_1), \ldots, Q_{t-n}(J_1)) \tag{3}$$

where:

Q – traffic flow at junctions, current and past values.

The single NN model combines two NNs into one in order to reduce the complexity and utilise the networks capability for finding mutual relations between the junctions flows.

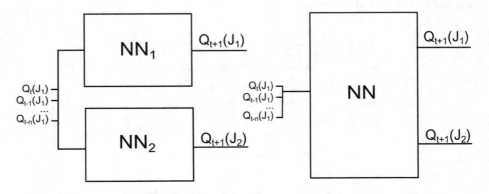

Fig. 2. Prediction models using NNs

There are two variables shaping the complexity of the models: traffic flow history and construction parameters of the incorporated NNs. Three neural network types are evaluated: multilayer perceptron, cascade forward network and deep learning network with autoencoders.

The length of traffic flow history, which is to be taken into account for forecasting is determined on the basis of variability of changes of flow values in time [2]. In [3] traffic flow time series are analysed for rating the length of periods of traffic changeability. The duration of traffic peaks, periods of traffic build up or slowdown are evaluated for finding a common period of changes. One hour period covers the characteristic changes of observed traffic flows. In this study five consecutive past 15 min traffic flow values are used as inputs, which amounts to a 75 min history of traffic flow.

The number of neurons of the network layers depends on the character and complexity of the changes of the input data. If the analysed data set has no distinct features the network structure must incorporate more neurons to map the detailed character of data. The optimal NN structure is obtained in an iterative manner in combination with a number of training sessions.

4 Validation of the Neural Network Designs

Reliable data sets form the basis for investigating the properties of neural networks. The validation of constructed networks requires a number of training and test sets. Thanks to the help of the local Traffic Control Centre there was practically no limit on the size of the data sets which were used for validating the design.

In the case of this control centre, vehicles counts are done in 5 min intervals and converted to average flow values over 15 min periods. In all over 15 thousand flow values, covering over 200 days of traffic observations are used to train and an appropriate number to test the networks.

The neural network structure is changed in a systematic way by incremental modifications of the number of layers, the number of neurons in the layers. The final change is done by combining the double network structure into one network with two outputs.

4.1 Data Sets for NN Validation

Data acquired from the measuring sites consists of vehicle counts registered in 5 min periods. The sites are situated on a traffic route (main road) leading to the city centre at two consecutive signalized junctions lying 800 m apart. The vicinity of the road is not a significant traffic generator. The sites are equipped with video detectors. The reliability of the vehicle counts was statistically checked by manual counting. The vehicle counts are summed up to obtain 15 min values and then converted to traffic flows. This reduces the scatter of data.

Work day traffic flow data is extracted from the vast data set of registered values. Three training sets of 4320 values (45 days × 96 flow values) are prepared and the same number of testing sets are extracted from the archives. In all over 15 thousand values are prepared for the validation of the NN design. The data sets comprise traffic parameters registered for the duration of over half a year.

4.2 NN Configurations

Validation of the prediction NNs is performed for the two basic configurations as presented in Fig. 2. The value of current traffic flow plus five consecutive past values of traffic flow are used as network inputs. The number of neurons in the layers varies from 1 to 36.

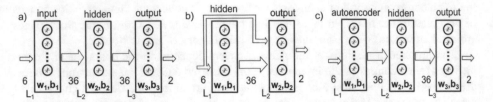

Fig. 3. NNs used in the design of the prediction models: (a) MLP, (b) CFN, (c) DLN

Figure 3 presents the optimal configurations of the networks obtained in the course of validation. Double and single network solutions are investigated.

The past values of traffic flow, which make up the inputs of the network describe the history of traffic changes. This amounts to 75 min.

5 Prediction Performance of the NNs

Systematic performance tests were carried out for the proposed set of optimisation variables. In all more than thirty network configurations were tested. An extensive set of results for analysis was obtained using the prepared test data sets.

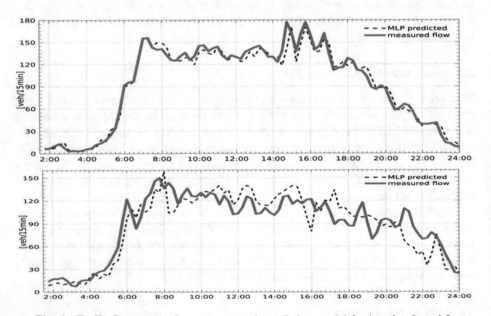

Fig. 4. Traffic flow graphs for a two network prediction model for junction J_1 and J_2

Accepted as representative are the prediction results of networks as in Fig. 3 that is MLP is configured (6-36-36-2) with one hidden layer, CFN uses one hidden layer with 36 neurons and the inputs are directed to the output layer together with the outputs of the hidden layer, the DLN uses one layer of autoencoders generating 20 values for a 36 neuron hidden layer. These configurations retain the characteristic error behaviour which is observed for the prediction task

Figure 4 illustrates the deviations of the forecast values from the measured traffic flow for prediction with separate NNs for each junction. In the case of the first junction J_1 there is a high compliance of the values, the calculated MAPE reached 5.3%. Discrepancies arise when the course of the flow values slowly changes its direction. This character of change is poorly mapped by the network. The predictions for the second junction J_2 differ to much a larger extent MAPE reaches 15%. Similar behaviour is observed as in the case of J_1, but the values of the differences between the graphs are higher. Only the MLP network based solution is presented.

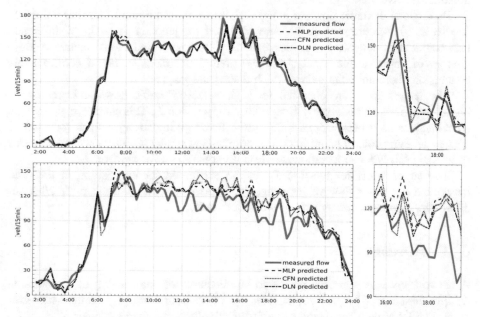

Fig. 5. Traffic flow graphs for a single network prediction solution for junction J_1 and J_2

Figure 5 presents the prediction results of a solution based on a single NN with two outputs. In this case the character of errors is slightly changed that is: slow changes of flow values in some periods generate abrupt changes (peaks) of predicted values. The predicted values are delayed relative to measurements. Enlarged views of characteristic runs are appended to the right of the graphs in Fig. 5.

A detailed evaluation of the networks performance is done using error measures. Mean absolute error (MAE) values are determined and compared with root mean square error (RMSE). The MAE and the RMSE are used together to diagnose the variation in the errors in the set of forecasts. The difference between these errors will indicate the significance of large errors. RMSE assigns higher weights to large differences and may signal a large content of outliers in the set of prediction values. The larger difference between them, the larger the discrepancy in the individual errors.

Table 1. Prediction errors of the NN model with a single network

NN configuration	J_1			J_2		
	RMSE [veh/15 min]	MAE [veh/15 min]	MAPE [%]	RMSE [veh/15 min]	MAE [veh/15 min]	MAPE [%]
MLP 6-36-36-2	8.1	6.0	5.1	17.5	14.1	14.0
CFN 6-36-2	8.1	6.1	5.3	18.6	14.8	14.4
DLN 6-20-36-2	8.0	6.0	5.1	16.7	13.4	13.3

Table 1 presents the errors of prediction for junction J_1 and J_2 respectively. Additionally a column of the mean absolute percentage errors (MAPE) is attached to illustrate relative values of errors.

Predictions of traffic flow for junction J_1 have about two times smaller values of RMSE and MAE errors than the ones obtained for junction J_2. A large difference between RMSE and MAE signifies that the NN based predictions contain a large number of outlier errors. Analysis of the prediction graphs in Fig. 5 confirms this conclusion, there are point values which distort the graph.

The largest errors are observed for high values of traffic flows and predictions higher than measurements prevail, especially for junction J_2. This characteristic may be advantageous as it over estimates the traffic flows giving stronger signs for need of change of traffic control parameters. The comparison results for junction J_1 in all cases of investigated NN configurations are less divergent than for junction J_2.

The larger errors for junction J_2 can be accounted for its position in the road network it has lower flow values which indicates that some traffic from J_1 does not arrive at J_2, perhaps drivers stop or use other routes.

6   Conclusions

A network growing approach proves to be adequate for optimisation of a prediction solution based on NNs. The investigated set of configurations give comparable error rates. The DLN with autoencoders performs slightly better than the other NNs.

A single NN can provide predictions for two consecutive junctions based on measured and historical traffic flow values of the first junction in sequence. Results prove that the design may be incorporated as a component of a prediction module in Intelligent Transport Systems.

The phenomena of poor mapping of slowly changing traffic flows, at high flow values, requires further detailed examination. This will be the subject of future research works.

The use of traffic data from adjacent junctions to obtain more accurate forecasting results will be further investigated.

Acknowledgments. The authors wish to thank ZIR-SSR Bytom for providing video detector data from the Traffic Control Centre Gliwice site.

References

1. Pamuła, T.: Classification and prediction of traffic flow based on real data using neural networks. Arch. Transp. **24**, 519–529 (2012)
2. Bernaś, M., Płaczek, B., Porwik, P., Pamuła, T.: Segmentation of vehicle detector data for improved k-nearest neighbours-based traffic flow prediction. IET Intell. Transp. Syst. **9**, 1–11 (2015)
3. Pamuła, T.: Traffic flow analysis based on the real data using neural networks. In: Mikulski, J. (ed.) Telematics in the Transport Environment. Communications in Computer and Information Science, vol. 329, pp. 364–371. Springer, New York (2012)
4. Xiaoying, L.: Prediction of traffic flow base on neural network. In: Intelligent Computation Technology and Automation, ICICTA 2009, pp. 374–377 (2009)
5. Vlahogianni, E.I., Karlaftis, M.G., Golias, J.C.: Optimized and meta-optimized neural networks for short-term traffic flow prediction: a genetic approach. Transp. Res. C **13**, 211–234 (2005)
6. Zhu, J.Z., Cao, J.X., Zhu, Y.: Traffic volume forecasting based on radial basis function neural network with the consideration of traffic flows at the adjacent intersections. Transp. Res. C **47**, 139–154 (2014)
7. Karlaftis, M.G., Vlahogianni, E.J.: Statistical methods versus neural networks in transportation research: differences, similiarities and some insights. Transp. Res. C **19**, 387–399 (2011)
8. Król, A.: The application of the artificial intelligence methods for planning of the development of the transportation network. In: Rafalski, L., Zofka, A. (eds.) 6th Transport Research Arena, TRA 2016. Transp. Res. Procedia **14**, 4532–4541. Elsevier (2016). ISSN 2352-1465
9. Vlahogianni, E.I., Karlaftis, M.G., Golias, J.C.: Short-term traffic forecasting: where we are and where we're going. Transp. Res. C **43**, 3–19 (2014)
10. Centiner, B.G., Sari, M., Borat, O.: A neural network based traffic-flow prediction model. Math. Comput. Appl. **15**, 269–278 (2010)
11. Kumar, K., Parida, M., Katiyar, V.K.: Short term traffic flow prediction for a non urban highway using artificial neural network. Procedia Soc. Behav. Sci. **104**, 755–764 (2013)
12. Lorenzo, M., Matteo, M.: OD matrices network estimation from link counts by neural networks. J. Transp. Syst. Eng. Inf. Technol. **13**(4), 84–92 (2013)
13. Park, J., Murphey, Y.L., McGee, R., Kristinsson, J.G., Kuang, M.L., Phillips, A.M.: Intelligent trip modeling for the prediction of an origin-destination traveling speed profile. IEEE Trans. Intell. Transp. Syst. **15**(3), 1039–1053 (2014)
14. Lv, Y., Duan, Y., Kang, W., Li, Z., Wang, F.Y.: Traffic flow prediction with big data: a deep learning approach. IEEE Trans. Intell. Transp. Syst. **16**(2), 865–873 (2015)
15. Yang, H.F., Dillon, T.S., Chen, Y.P.P.: Optimized structure of the traffic flow forecasting model with a deep learning approach. IEEE Trans. Neural Netw. Learn. Syst. (2016). https://doi.org/10.1109/TNNLS.2016.2574840

Advanced Data Collecting, Analysis and Intelligent Support for Decision Making

Data-Driven Transport Policy in Cities: A Literature Review and Implications for Future Developments

Anna Urbanek(✉)

Faculty of Economics, University of Economics in Katowice, Katowice, Poland
anna.urbanek@ue.katowice.pl

Abstract. The skill of analysis of big data provided by the ICT and the skill of using this information in the decision-making process become crucial elements of the increase in competitiveness and effectiveness in all sectors of economy, in particular in the transport sector, in the field of mobility management in cities. The data-driven transport policy is also one of key pillars of a smart city concept. The paper discusses the importance of data-driven decision making in the common transport policy of the European Union. The paper reviews also the carried out so far studies on the transport behaviour, utilising passive data streams. Finally, the paper discusses main challenges facing researchers and public transport authorities related to the use of passive data streams for the needs of studies in the future.

Keywords: Big data · Mobility · Transport policy · Travel behaviour
Passive data stream · Transport demand · Automated fare collection system
Smart cards · Urban transport · Public transport · Smart city · Smart governance

1 Introduction

The knowledge-based economy is a relatively new concept, which became popular after the publication in 1996 of an OECD report entitled 'The knowledge-based economy'. According to the OECD definition the knowledge-based economies are such, which are based on direct use of knowledge and information for more effective functioning and improvement to the quality of life [1]. Together with an extremely dynamic progress in the ICT development in each area of life, which we have been witnessing in particular in the last decade, the original concept of knowledge-based economies becomes even more important and gains a new dimension. Because recent 10–15 years of global information society development have shown that just the information becomes today, parallel to classical resources, i.e. land, labour, and capital, a basic resource of each organisation, as well as one of basic economic categories.

The big data analysis is increasingly important not only in the private sector, but also in the public sector, where process optimisation and more effective management of public assets are especially necessary. The skill of information analysis and its use in the decision making process becomes one of key competitiveness elements in all sectors of economy, including in particular the transport sector, in the field of mobility

© Springer Nature Switzerland AG 2019
G. Sierpiński (Ed.): TSTP 2018, AISC 844, pp. 61–74, 2019.
https://doi.org/10.1007/978-3-319-99477-2_6

management in cities. Transport is one of main factors enabling the city functioning and development, and cities in turn are the drivers of the economy. In European Union states city residents constitute approx. 75% of citizens [2], and this figure is forecast to grow to approx. 84% by 2050 [3]. In 2016 there were 31 megacities worldwide, that is cities with a population exceeding 10 million, and their number is forecast to increase to 41 in 2030 [4]. The increasing population of cities and at the same time the expectation of good living and travelling conditions in cities are a challenge for the urban transport authorities. A quantitative development of transport in cities is limited not only by the urban space, the transport network capacity, but also by possibilities to finance investments from public funds. Hence the role of information about the demand for transport services and about the transport behaviour is so great, as it becomes the basis to make decisions in the field of pursuing a more effective use of the possessed resources.

The paper is aimed at reviewing the literature of the subject from the point of view of big data use in the studies on transport behaviour and in the process of mobility management in cities. The first part of the paper discusses the importance of data-driven decision making in the transport policy of the European Union. Part 2 reviews the studies carried out so far, utilising the big data, which sources include systems of automated fare collection in the public transport, GPS, and mobile phones. The third part of paper discusses challenges facing researchers and public transport authorities in relation to the use of passive data streams for the research needs in the future.

2   Data-Driven Decision Making: EU Transport Policy Perspective

The European Union documents more and more often refer to data-driven economies or data economies, as a foundation of knowledge economies in the future. Data economies are such, in which all market participants cooperate to ensure access to the information and data, from the used IT systems, which are to improve the quality of life via more effective use of resources, processes and decisions optimisation, as well as to accelerate the innovation and to support the forecasting processes [5].

The building of European data economy is a part of Digital Single Market Strategy adopted in 2015, in which the maximum utilisation of potential resulting from the digitisation for the economies and societies development is one of main objectives [6]. The core of such policy includes data, and basically this, what we call today the big data, i.e. big data sets, very diversified in terms of structure, kind, and type, frequently available in real time or close to it [7]. The big data is generated by all devices, which have an indirect or direct capability to collect, process, or exchange the data through computer networks.

The big data analysis is a source of numerous economic and social benefits, because it provides knowledge within such scope, time, and accuracy, as was not achievable so far using other standard research methods, allowing to make both current and strategic decisions based on the data from many sources, and thereby to evaluate events in a comprehensive way. Studies carried out by a team of researchers from the Massachusetts Institute of Technology and the University of Pennsylvania show that companies making decisions on the basis of data analysis based on e.g. business intelligence systems achieve approx. 5–6% higher output and productivity than those, which base on traditional input data and a standard use of IT systems [8].

A more and more common use of ICT created a possibility to analyse the big data almost in all sectors of economy. As the 2016 Eurostat data shows 10% of all enterprises in the European Union were carrying out big data analyses for the needs of their business. Table 1 presents the share of all enterprises and those in the transport and warehousing sector, which analysed the big data in 2016. The data in Table 1 take into account only enterprises employing at least 10 persons and do not consider the financial sector, for which the big data analysis is now one of main business pillars. Malta and the Netherlands are the UE states, in which the biggest number of enterprises, because as many as 19%, analyse the big data.

Table 1. Enterprises analysing big data (outside the financial sector, employing at least 10 persons) and main sources of big data acquisition in the European Union and selected EU countries in 2016 [2].

Country/group	Enterprises analysing big data (% of all enterprises)	Sensors, smart devices (e.g. using M2M communication, RFID tags, etc.)	Geolocation data from portable devices (e.g. GPS, smartphones, mobile phones)	Data generated by social media, blogs, or websites used for information exchange	Other sources
		% of big data analysing enterprises			
Malta	19	43	32	64	29
Netherlands	19	49	40	52	18
UK	15	22	30	67	32
Lithuania	12	49	59	51	51
France	11	29	62	32	10
Slovakia	11	40	44	35	28
UE 28	10	33	47	45	25
Czech Republic	9	46	53	26	28
Italy	9	33	36	29	39
Spain	8	34	49	51	20
Hungary	7	48	57	41	20
Poland	6	30	67	36	4
In the transportation and storage sector					
Netherlands	26	56	77	20	14
France	25	20	92	n.a.	4
Malta	22	27	50	59	13
UE 28	15	29	83	13	12
Italy	15	34	62	6	23
Spain	15	29	87	12	12

Table 1. (*continued*)

Country/group	Enterprises analysing big data (% of all enterprises)	Sensors, smart devices (e.g. using M2M communication, RFID tags, etc.)	Geolocation data from portable devices (e.g. GPS, smartphones, mobile phones)	Data generated by social media, blogs, or websites used for information exchange	Other sources
		% of big data analysing enterprises			
UK	14	28	73	21	8
Lithuania	14	37	84	48	48
Czech Republic	14	38	93	9	11
Hungary	10	34	88	6	11
Poland	10	20	94	5	0
Slovakia	6	63	63	7	8

As the table shows in 2016 in the EU 15% of enterprises in this sector used analyses of big data sets in their business. In such countries as the Netherlands or France such analyses are carried out by every fourth enterprise, in Poland by every tenth. The big data sources in the transport sector are dominated by the geolocation data originating from various mobile devices transmitting the information wireless, e.g. via mobile networks, GPS, etc. The transport and storage sector, after the ICT sector (24% of all enterprises in the EU28) and electricity, gas, and water suppliers (16% of all enterprises in the EU28), is the sector of economy, in which the demand for the big data analysis is the highest, as shown by the data.

In the European Commission (EC) guidelines the transport, and in particular the transport in cities is indicated as this area, in which the use of knowledge resulting from the ICT application and from the analysis of big data provided by this technology, is a crucial element of effectiveness increase. In the European Commission communique 'Towards a thriving data-driven economy' integrated transport management systems in cities and regions are mentioned as those of strategic importance in building the EU data economy [9]. Also the European Union transport policy is consistent from this point of view. All strategic documents determining the EU transport policy objectives indicate, that the improvement in conditions of movement in cities and the urban mobility planning are not possible without advanced technologies related to the ITS implementation and to the usage of data provided by them (Table 2).

In the 'Together towards competitive and resource-efficient urban mobility' communique the European Commission draws attention to the fact that smart technologies and ITS systems play a key role in developing city policies [10]. It is also not possible to pass over a number of regulations, including also decisions and directives concerning the ITS deployment in the European Union, related to the performance of the 'Action Plan for the Deployment of Intelligent Transport Systems in Europe' [11].

Without advanced information and communication technologies and also without the analysis and use of data delivered by such systems in the decision making process it is also not possible to refer to a smart city concept. The main components of a smart city most frequently include: smart economy, smart mobility, smart environment, smart people, smart living, and smart governance [16]. There are many definitions of a smart city, but the essence of smart city vision consists in using the knowledge, organisational solutions, and first of all modern ICT technologies for significant improvement in the city functioning and in the quality of life in cities [17–19]. It is the information and its proper use that is in the centre of this concept, and each of smart city dimensions, in particular economy, environment, mobility, and governance, is based on data-driven decision making and as a result on data-driven urban policy making.

Table 2. Examples of European Commission documents determining the transport policy objectives, recommending actions in the field of big data acquisition, analysis, and use in cities and regions [10, 12–15].

Document	Examples of recommended actions in the field of big data acquisition, analysis, and use in cities and regions
EC. Green paper – Towards a new culture for urban mobility, COM(2007)551 final	The need to reduce the congestion in cities thanks to co-modality and intelligent and adaptive traffic management systems Efficient management of urban mobility thanks to Intelligent Transport Systems (ITS) applications: flexible and smart charging system and interoperable multi-modal trip information
EC. Action Plan on Urban Mobility, COM (2009) 490 final	The need to improve data collection for urban transport and mobility and the necessity to exchange information between various stakeholders, and primarily between local and regional authorities across the world, which face similar challenges in the field of mobility management Optimising urban mobility and pursuing sustainable urban mobility due to ITS applications for urban mobility (i.e.: e-ticketing, traffic management, travel information, access regulation and demand management)

(continued)

Table 2. (*continued*)

Document	Examples of recommended actions in the field of big data acquisition, analysis, and use in cities and regions
EC. White Paper - Roadmap to a Single European Transport Area – Towards a competitive resource efficient transport system, COM(2011)144 final	Deployment of smart mobility systems in each mode of transport (i.e. SESAR, ERTMS, RIS, SafeSeaNet). Deployment of "integrated transport management and information systems, facilitating smart mobility services, traffic management for improved use of infrastructure and vehicles, and real-time information systems to track and trace freight and to manage freight flows; passenger/travel information, booking and payment systems" Interoperable and intelligent infrastructure that ensure monitoring and communication between different elements of transport system
EC. Together towards competitive and resource-efficient urban mobility, COM (2013)913 final	The ITS development is a condition for the optimisation of the existing infrastructure use and for the source-efficient urban mobility "Member States should consider setting-up interoperable multimodal datasets gathering all information about urban mobility" and the EC "will take forward work on supplementing the existing legislation on access to traffic and travel data"
EC. Europe on the move – an agenda for a socially fair transition towards clean competitive and connected mobility for all, COM(2017)283 final	"The deployment of (..) interconnected and cross-border infrastructures and interoperable digital services (5G coverage, data networks, cooperative ITS (…)"

3   Big Data for Mobility Studies – a Literature Review

Traditional methods of information collection for the needs of studies on transport behaviour and mobility related analyses are based mainly on manual information gathering by a properly trained team working in the field. The following can be primarily mentioned among most frequently used methods:

- questionnaire surveys of households, origin-destination surveys, or travel journals,
- surveys of passenger flows in the form of vehicle occupancy surveys, cordon line surveys etc.

Traditional research methods, based on the use of a questionnaire form, obviously have some advantages. First of all they allow to obtain more detailed information, but also allow to perform surveys in relation to respondents characteristics such as sex, level of education, performed profession, or the level of wealth. Nevertheless, traditional research methods feature also numerous drawbacks and limitations. Primarily, these are relatively expensive methods and very demanding in terms of organisation, which results in their relatively rare performance. Depending on the city and scope, such surveys are carried out every two, and even 5 or 10 years. Moreover, traditional research methods allow to learn the reality only in a certain selected representative period, during which such the survey is performed. As a result they do not provide the knowledge about the demand for services or about the transport behaviour in nontypical periods, which result from e.g. unexpected events or from seasonal fluctuations. In addition, the use of traditional research methods is always related to a very high risk of error and to limitations due to the used statistical methods (sample selection and size, sample randomness, sample representativeness, etc.) [20, 21].

All that makes that traditional research methods are a tool of very limited usefulness in the process of decision making in the business practice. Therefore increasingly often and on a larger and larger scale so-called passive data streams are used in the mobility surveys, i.e. big data generated by all devices having a capacity to collect and exchange the data via computer networks, and to some extent without the user awareness [22]. They enable the information analysis on a huge and so far unavailable scale in any period of time. The mobility surveys use now primarily all the information originating from the automated fare collection systems (AFC), which in majority of cities worldwide are based on smart cards and mobile ticketing [22, 23].

The literature of the subject provides numerous examples of the use of data from smart cards for the needs of mobility analyses in specific cities. Table 3 specifies examples of mobility studies using the data from smart cards in specific cities worldwide and the main results of studies, being the basis for decisions taken in the field of public transport planning and management, and also of pricing policy.

Table 3. Examples of mobility studies using the data from the smart cards usage [20, 24–33].

Author(s)	Studied area/city	Main results of the study
Bagchi and White [24]	Southport, Merseyside, Bradford (UK)	Mobility analyses for individual passenger groups Turnover analysis Study of the smart cards use potential for the needs of public transport management and planning
Utsunomiya et al. [25]	Chicago, USA (Chicago Transit Authority)	Mobility analyses for individual passenger groups Better understanding of passengers behaviour Public transport offer adaptation to passenger needs Demand forecasting

(*continued*)

Table 3. (*continued*)

Author(s)	Studied area/city	Main results of the study
Agard, Morency and Trépanier [26]	Gatineau, Quebec, Canada	Separation of market segments among public transport users Transport behaviour analysis linked to card type
Barry et al. [28] Barry et al. [27]	New York, USA (Metropolitan Transit Authority's New York City Transit)	Recognition of travel patterns Origin-destination (OD) matrices Network load profiles Formulation of conclusions to improve planning and management
Liu et al. [29]	Shenzen, China	Recognition of spatial and temporal mobility patterns Network load visualization Conclusions for the public transport network design and optimization
Wang, Attanucci and Wilson [30]	London, UK (Oyster and iBus)	Origin-destination (OD) matrices Network Load/Flow Profiles Interchange Time Analysis
Munizaga and Palma [31]	Santiago, Chile (Transantiago)	Origin-destination (OD) matrices Conclusions to improve multimodal network planning and management
Lee and Hickman [32]	Metro Transit in the Minneapolis/St. Paul metropolitan area	Origin-destination (OD) matrices
Ma et al. [33]	Beijing, China	Spatiotemporal analysis of commuters and non-commuters behaviour Recognition of travel patterns Conclusions for the public transport network design and optimization
Briand et al. [20]	Gatineau, Quebec, Canada	Evolution of passenger behaviour in time Spatiotemporal analysis of mobility linked to the fare types

The data from smart cards usage is now one of best big data sources for various analyses and studies related to mobility. Mainly because of [23]:

- very broad functionality of smart cards in public transport (carrier of different ticket types for various means of transport and various organisers),
- possibility to identify and record all transactions performed by passengers (they can be an e-money carrier and operate as payment cards, possibility to identify exact vehicle entry and exit stops, to use various fares both for single-travel and season tickets),
- possibilities of personalising and assigning specific entitlements, e.g. concessions, of linking the data with specific passenger demographic features,

- access to the data (usually such data is held and managed by a public transport authority or an AFC system entity/operator, with whom they have a contract).

Therefore the studies using just the information from smart cards are the most widespread and most frequently used for the needs of connections network planning and management. Pelletier, Trépanier and Morency already in 2011 reviewed such studies from the point of view of data from smart cards usage for the needs of strategic, tactical, and operational planning [34].

Very often the smart card data is used in connection with the GPS devices data, which are an integral part of vehicles already in almost all public transport systems worldwide. The GPS data allow to control the performance of transport work, the correctness of journeys performance, and the assessment of provided services quality (speed, delays, etc.). Linked with the smart cards data it is an extremely precise and necessary tool, supporting public transport authorities in the planning and management process. The study carried out by Geschwender, Munizaga and Simonetti in Santiago, Chile [35] can be an example of such studies and cooperation between a local transport authority and researchers.

Also the data from mobile phones can be the source of data for mobility related studies. More and more such analysis are carried out in recent years. Table 4 presents examples of analyses using the data from mobile phones from specific areas or cities, including the main analysis results.

Table 4. Review of studies on mobile phones in urban mobility [36–40].

Author(s)	Studied area/city	Main results of the study
González et al. [36]	–	Mobility patterns Spatiotemporal analysis of mobility The shapes of human trajectories
Calabrese et al. [37]	The Boston Metropolitan Area	Spatial distribution of daily mobility Individual mobility analysis (two measures the average daily total trip length and the average daily VKT per vehicle)
Iqbal et al. [38]	Dhaka, Bangladesh	Origin-destination (OD) matrices Traffic flows on network
Alexander et al. [39]	The Boston Metropolitan Area	Origin-destination (OD) matrices Spatial distribution of intra-town and inter-town flows
Geurs et al. [40]	the Dutch Mobile Mobility Panel (MoveSmarter App)	Modal shift Trip length and travel time distribution Daily average number of trips

The mobile phone data has a huge potential of application in studies on mobility due to the fact, that today almost everybody has a telephone with him/her. In addition, smartphones are equipped with numerous sensors, such as GPS, accelerometer, gyroscope, Wi-Fi etc., the information from which linked with the data on telephone location can be a source of huge knowledge about the transport behaviour. That means that the mobile phone data can be helpful in studies on mobility of all city residents or persons staying in a given city or in a given area, and not only of public transport users, as it is the case of smart cards. Hence the data acquired from smart phones has a huge potential in studies on car users, pedestrians, and bicycle riders. Obviously a significant drawback of mobile phone data is the difficulty in data acquisition (the necessity to acquire the data from mobile networks operators), and also in practice - due to the personal data protection - impossibility to link this data with users socio-demographic data.

4 Implications for Future Solutions

None of hitherto known and used research methods can provide such knowledge, as the big data analysis. It is the big data analysis that is the future of studies on transport behaviour and of mobility related analyses. Full utilisation of potential resulting from the big data analysis for the needs of mobility related studies in cities will force development of new forms of cooperation and partnership in the field of data sharing, primarily between the private and public sector [41, 42]. The private sector already now has available huge data sets, which are used for own business needs. To manage mobility in smart cities in the future it will be necessary to aggregate and analyse the data not only from transport companies and ITS systems in the city, but also the information from companies operating in other sectors, e.g. to forecast the demand for transport services. It is possible to presume that in the near future it will become necessary to introduce an obligation of making the data available for transport authorities by all entities, in particular by:

- mobile networks operators and owners of smartphone applications,
- freight transport operators to manage the flows of goods in cities [43],
- taxi, car rental, carsharing, and carpooling companies,
- private companies operating in the passenger transport sector.

Moreover, it will be also necessary to intensify actions to create a new quality in the field of access to data for citizens and to implement open data policy on various levels - local, regional, national and international. Already today this is one of European Union policy objectives under the Digital Single Market Strategy. This data can become a source for software developers, e.g. creating travel applications for smartphones, which can be used by city residents. The Transport for London open data programme [44] is an example of such solution operating successfully already today.

Full utilisation of potential resulting from the big data analysis will require to make closer the cooperation between the academic sector and researchers and the economic practice, in particular transport authorities in cities. In the past the data acquisition was the main problem of mobility and transport behaviour analyses. Today the vastness and diversification of received data make that the problem consists in their categorisation and analysis, so as to make them useful in decision making. As a result it is possible to presume that it will force also certain organisational changes and changes in the model of public entities functioning.

In this context it should be also emphasised, that one of greatest challenges facing policy makers, the sector of science and the private sector, consists in creating and effectively implementing standards and harmonisation rules for the collected data, as well as developing common platforms for their exchange, so that they could be freely exchanged and used by various entities, not only locally or regionally, but also on the international scale, e.g. for comparative studies [45]. The first actions in this field have already been taken by the European Commission under the INSPIRE Directive [46], nevertheless the IT systems interoperability is not an easy task and this process must proceed in stages.

In addition, the new big data sharing models require great caution in the field of privacy and personal data safety protection. That will require establishing procedures related to making the mobility related data anonymous. A possibility to collect precise data on the demand for transport services, which is beyond any doubt the greatest advantage of ICT, on the other hand is a significant element raising social fears related to the protection of users' privacy and to the aim of information use. The studies carried out so far show that these fears frequently decide also about such systems acceptance by the society [34, 44].

5 Conclusions

The progressing urbanisation and the striving for an increase in the effectiveness of provided services make that the role of using the information brought by big data in the process of mobility management in cities has been growing. The data-driven transport policy becomes a condition for effective use of resources and a tool to shape sustainable transport development in cities. It is also one of pillars for smart cities building in the future.

The ICT use and the big data analysis in the process of mobility management is also one of main pillars of the European Union common transport policy. Although this is not a new issue in the EU policy, but in fact the last decade was related to significant intensification of work in this field.

Studies carried out worldwide so far show that the potential of passive data streams application for mobility related analyses is huge. Because of the research and analytical methods developed so far it is possible to analyse the travel behaviour in various cross-sections and to formulate conclusions for the needs of public transport network planning and optimising. Moreover, the mobile phone data create a possibility to study the transport behaviour not only of public transport users, but also that of pedestrians, bicycle riders, and car users.

The big data created a possibility to carry out studies on the hitherto unavailable scale of observations number, time, and precision. However, the big data analysis and translation of the available data vastness into specific decisions in the economic practice is not an easy task, because it requires application of sophisticated analytical and research methods. The carried out analysis shows that the cooperation of the academic sector with the transport policy entities on various levels is a condition for full utilisation of the big data potential in the process of mobility management in cities in the future.

References

1. The knowledge-based economy: OECD, Paris (1996). https://www.oecd.org/sti/sci-tech/1913021.pdf
2. Eurostat Database. http://ec.europa.eu/eurostat
3. United Nations, Department of Economic and Social Affairs, Population Division: World Urbanization Prospects: The 2014 Revision, Highlights (ST/ESA/SER.A/352) (2014)
4. United Nations, Department of Economic and Social Affairs, Population Division: The World's Cities in 2016: Data booklet (2016). http://www.un.org/en/development/desa/population/publications/pdf/urbanization/the_worlds_cities_in_2016_data_booklet.pdf
5. European Commission: Building a European Data Economy, COM/2017/09/final, Brussels. http://eur-lex.europa.eu. Accessed 10 Jan 2017
6. European Commission: A Digital Single Market Strategy for Europe, COM/2015/0192 final, Brussels. http://eur-lex.europa.eu. Accessed 6 May 2015
7. Einav, L., Levin, J.D.: The data revolution and economic analysis. Working paper series no. 19035, National Bureau of Economic Research, Cambridge (2013)
8. Brynjolfsson, E., Hitt, L.M., Kim, H.H.: Strength in Numbers: How Does Data-Driven Decision-Making Affect Firm Performance? https://ssrn.com/abstract=1819486. Accessed 22 Apr 2011
9. European Commission: Towards a thriving data-driven economy, COM/2014/0442/final, Brussels. http://eur-lex.europa.eu. Accessed 2 July 2014
10. European Commission: Together towards competitive and resource-efficient urban mobility, COM/2013 913 final, Brussels. http://eur-lex.europa.eu. Accessed 17 Dec 2013
11. European Commission. Action Plan for the Deployment of Intelligent Transport Systems in Europe, COM/2008/0886 final, Brussels. http://eur-lex.europa.eu. Accessed 16 Dec 2008
12. European Commission: Green paper – Towards a new culture for urban mobility. COM (2007) 551 final, Brussels. http://eur-lex.europa.eu. Accessed 25 Sep 2007
13. European Commission: Action Plan on Urban Mobility, COM (2009) 0490 final, Brussels. http://eur-lex.europa.eu. Accessed 30 Sep 2009
14. European Commission: White Paper - Roadmap to a Single European Transport Area – Towards a competitive resource efficient transport system, COM(2011) 144 final, Brussels. http://eur-lex.europa.eu. Accessed 28 Mar 2011
15. European Commission: Europe on the move – an agenda for a socially fair transition towards clean competitive and connected mobility for all, COM(2017) 283 final, Brussels. http://eur-lex.europa.eu. Accessed 31 May 2017
16. European Parliament: Mapping Smart Cities in the EU (2014). http://www.smartcities.at/assets/Publikationen/Weitere-Publikationen-zum-Thema/mappingsmartcities.pdf
17. Tomanek, R.: Sustainable mobility in smart metropolis. In: Brdulak, A., Brdulak, H. (eds.) Happy City - How to Plan and Create the Best Liveable Area for the People. EcoProduction (Environmental Issues in Logistics and Manufacturing), pp. 3–16. Springer, Cham (2017)
18. Nam, T., Pardo, T.A.: Smart city as urban innovation: focusing on management, policy, and context. In: Proceedings of the 5th International Conference on Theory and Practice of Electronic Governance, pp. 185–194. Tallinn, Estonia (2011)
19. Gil-Garcia, J.R., Zhang, J., Puron-Cid, G.: Conceptualizing smartness in government: an integrative and multi-dimensional view. Gov. Inf. Q. **33**(3), 524–534 (2016). https://doi.org/10.1016/j.giq.2016.03.002
20. Briand, A.S., Côme, E., Trépanier, M., Oukhellou, L.: Analyzing year-to-year changes in public transport passenger behaviour using smart card data. Transp. Res. Part C: Emerg. Technol. **79**, 274–289 (2017). https://doi.org/10.1016/j.trc.2017.03.021

21. Clifton, K.J., Handy, S.L.: Qualitative methods in travel behavior research. In: Jones, P., Stopher, P.R. (eds.) Transport Survey Quality and Innovation, pp. 283–302. Emerald, Bingley (2003)
22. Trépanier, M., Tamamoto, T.: Workshop synthesis: system based passive data streams systems; smart cards, phone data, GPS. Transp. Res. Procedia **11**, 340–349 (2015)
23. Urbanek, A.: Big data – a challenge for urban transport managers. Commun. Sci. Lett. Univ. Zilina **19**(2), 36–42 (2017)
24. Bagchi, M., White, P.R.: The potential of public transport smart card data. Transp. Policy **12** (5), 464–474 (2005)
25. Utsunomiya, M., Attanucci, J., Wilson, N.H.: Potential uses of transit smart card registration and transaction data to improve transit planning. Transp. Res. Rec.: J. Transp. Res. Board **1971**, 119–126 (2006). Transportation Research Board of the National Academies, Washington, D.C.
26. Agard, B., Morency, C., Trépanier, M.: Mining public transport user behaviour from smart card data. IFAC Proc. Vol. **39**, 399–404 (2006). https://doi.org/10.3182/20060517-3-FR-2903.00211
27. Barry, J., Freiner, R., Slavin, H.: Use of entry-only automatic fare collection data to estimate linked transit trips in New York City. Transp. Res. Rec. **2112**, 53–61 (2009)
28. Barry, J., Newhouser, R., Rahbee, A., Sayeda, S.: Origin and destination estimation in New York City with automated fare system data. Transp. Res. Rec.: J. Transp. Res. Board **1817**, 183–187 (2002). Transportation Research Board of the National Academies, Washington, D.C.
29. Liu, L., Hou, A., Biderman, A., Ratti, C., Chen, J.: Understanding individual and collective mobility patterns from smart card records: a case study in Shenzhen. In: 12th International IEEE Conference on Intelligent Transportation Systems, ITSC, pp. 842–847 (2009)
30. Wang, W., Attanucci, J., Wilson, N.: Bus passenger origin-destination estimation and related analyses using automated data collection systems. J. Public Transp. **14**(4), 131–150 (2011). https://doi.org/10.5038/2375-0901.14.4.7
31. Munizaga, M.A., Palma, C.: Estimation of a disaggregate multimodal public transport origin-destination matrix from passive Smart card data from Santiago, Chile. Transp. Res. Part C **24C**(12), 9–18 (2012)
32. Lee, S.G., Hickman, M.: Trip purpose inference using automated fare collection data. Public Transp. **6**, 1–20 (2014). https://doi.org/10.1007/s12469-013-0077-5
33. Ma, X., Liu, C., Wen, H., Wang, Y., Wu, Y.J.: Understanding commuting patterns using transit smart card data. J. Transp. Geogr. **58**, 135–145 (2017)
34. Pelletier, M.-P., Trépanier, M., Morency, C.: Smart card data use in public transit: a literature review. Transp. Res. Part C **19**(4), 557–568 (2011)
35. Geschwender, A., Munizaga, M., Simonetti, C.: Using smart card and GPS data for policy and planning: the case of transantiago. Res. Transp. Econ. **59**, 242–249 (2016). https://doi.org/10.1016/j.retrec.2016.05.004
36. González, M.C., Hidalgo, C.A., Barabási, A.-L.: Understanding individual human mobility patterns. Nature **453**, 779–782 (2008). https://doi.org/10.1038/nature06958
37. Calabrese, F., Diao, M., Di Lorenzo, G., Ferreira Jr., J., Ratti, C.: Understanding individual mobility patterns from urban sensing data: a mobile phone trace example. Transp. Res. Procedia Part C **26**, 301–313 (2013). https://doi.org/10.1016/j.trc.2012.09.009
38. Iqbal, M.S., Choudhury, C.F., Wang, P., Gonzáles, M.C.: Development of origin–destination matrices using mobile phone call data. Transp. Res. Procedia Part C **40**, 63–74 (2014). https://doi.org/10.1016/j.trc.2014.01.002

39. Alexander, L., Jiang, S., Murga, M., Gonzáles, M.C.: Origin-destination trips purpose and time of day interfered from mobile phone data. Transp. Res. Procedia Part C **58**, 240–250 (2015). https://doi.org/10.1016/j.trc.2015.02.018
40. Geurs, K.T., Thomas, T., Bijlsma, M., Douhou, S.: Automatic trip and mode detection with move smarter: first results from the Dutch mobile mobility panel. Transp. Res. Procedia **11**, 247–262 (2015). https://doi.org/10.1016/j.trpro.2015.12.022
41. International Transport Forum OECD: Corporate Partnership Board Report, Data-driven Transport Policy, Paris (2016)
42. International Transport Forum OECD: Corporate Partnership Board Report, Big data and Transport: Understanding and assessing options, Paris (2015)
43. Taniguchi, E.: Concepts of city logistics for sustainable and liveable cities. Procedia Soc. Behav. Sci. **151**, 310–317 (2014). https://doi.org/10.1016/j.sbspro.2014.10.029
44. Urbanek, A.: Automated fare collection systems based on check-in and check-out-premises of implementation in urban public transport. Arch. Transp. Syst. Telemat. **10**(3), 40–45 (2017)
45. Steenberghen, T., Pourbaix, J., Moulin, A., Bamps, C., Keijers, S.: Study on Harmonised Collection of European Data and Statistics in the Field of Urban Mobility. MOVE/B4/196-2/2010, Final report 24 May 2013, SADL KU Leuven and UITP (2013). https://ec.europa.eu/transport/sites/transport/files/themes/urban/studies/doc/2013-05-harmonised-collection-data-and-statistics-urban-transport.pdf
46. Directive 2007/2/EC of the European Parliament and of the Council of 14 March 2007 establishing an Infrastructure for Spatial Information in the European Community (INSPIRE). Official Journal of the European Union L 108, pp. 1–14, 25 April 2007. http://eur-lex.europa.eu

Using a Crowdsourcing Tool to Collect Data on the Travel Behaviour and Needs of Individuals with Reduced Mobility

Katarzyna Nosal Hoy[1(✉)] and Sylwia Rogala[2]

[1] Cracow University of Technology, Kraków, Poland
`knosal@pk.edu.pl`
[2] Kraków Municipal Infrastructure and Transport Board, Kraków, Poland
`srogala@zikit.krakow.pl`

Abstract. Sustainable mobility planning focuses on addressing the travel needs of city residents and improving their quality of life. The process of developing dedicated solutions relies on various tools of citizen participation, including particularly sensitive social groups, such as people with reduced mobility. The objective of the article is to present selected findings of a study conducted with the use of a web-based crowdsourcing tool to collect data on the travel behaviour and needs of citizens with reduced mobility in Kraków. These concerned aspects such as preferred means of transportation, reasons for the choice, and the evaluation of proposed solutions aimed at improving travel conditions. The final results include maps of trouble spots around the city, drawn up on the basis of spatial data generated by the respondents. The tool made it possible to analyze the travel behaviour of the study group and estimate the significance of various proposed solutions aimed at improving their travel conditions. The unique value of the study resides in the fact that it also examines the needs of child caregivers, a social group which has not yet been thoroughly studied.

Keywords: Sustainable mobility · Public participation
People with reduced mobility · Crowdsourcing

1 Introduction

Recently championed by the European Commission, Sustainable Urban Mobility Plans (SUMP) take a more sustainable and integrative approach to the challenges and problems of transport in urban areas, focusing on the needs and the quality of life of their residents [1, 2]. One of the central objectives of SUMPs is to guarantee that all residents enjoy access to the destinations of their choice, as well as to improve travel safety and comfort, especially on sustainable means of transportation [3, 4]. Since the plans are meant to help address the mobility needs of local communities, they are characterized by a participatory approach, where citizens and stakeholders are involved at every stage of the process, from early planning to final implementation [5]. Public participation makes it possible to understand how specific issues are perceived by local residents; communities can provide valuable information for the authorities [6], which helps increase the store of knowledge about transport supply and mobility demand [7].

© Springer Nature Switzerland AG 2019
G. Sierpiński (Ed.): TSTP 2018, AISC 844, pp. 75–84, 2019.
https://doi.org/10.1007/978-3-319-99477-2_7

Public involvement also allows to take account of the proposals, ideas, and fears of local communities [6]; this ensures that the proposed solutions will better correspond to their expectations, translating into greater social acceptance and reducing the risk of social protest [8].

Sustainable mobility plans must take into account the interests of all users, including particularly sensitive groups, such as individuals with reduced mobility [7, 9]. The latter are understood as all those who experience difficulties moving around environments that have not been specifically adapted to their needs [10], such as individuals in wheelchairs, people with limb problems and walking difficulties, people who are blind or visually impaired, deaf or hard of hearing, or suffer from various sensory, psychological, or intellectual disabilities [9, 11, 12]. Individual features related to the field of perception, such as communication problems affecting, e.g. those who do not, or barely, understand written and/or spoken language (including foreigners who do not speak the language of the place) are also associated with reduced mobility [11, 13]. The category also covers people with children in baby carriages, those carrying large or heavy luggage, seniors, pregnant women, and individuals of low stature (including children) [10, 13]. Considering the large number of possible temporary limitations (e.g. related to childcare or convalescence) and sporadic occurrences (e.g. moving around with a heavy or oversized luggage, visiting a foreign country), most people will experience reduced mobility at some time in their lives.

According to experts' estimations more than one third of the European population are people with reduced mobility, who face barriers when walking, biking and using public transport service [13]. These people have to depend on cars or specialized mobility services on many of their trips [13]. In many cases everyday services such as workplaces, shops, kindergartens, health centres, sports and leisure facilities are not easily accessible for pedestrians as well as bike and public transport users [14]. It is thus essential to take measures to improve their accessibility for those who walk, cycle, or use public transport; in doing so, the specific needs of people with reduced mobility should be addressed to ensure that their travel rights are respected on a par with those of other citizens [10]. Importantly, such an approach has various benefits across the table, since the measures help all residents and visitors to move around more freely and shape a less car-dependent lifestyle [13].

Cities looking to involve people with reduced mobility in sustainable mobility planning have a number of public participation tools to choose from [7]. However, important information on travel behaviour, barriers to mobility, and the needs of sensitive groups is usually gleaned from surveys, workshops, and focus groups [9, 15, 16]. These usually centre on seniors and people with disabilities (in wheelchairs) [12]. In contrast, the recent growth of online crowdsourcing instruments [17] has made it possible to tap the potential of the "crowd" to gather the data later used in the process of introducing dedicated solutions [7] to improve mobility, including among sensitive groups [12, 18, 19].

The objective of the current paper is to present the results of a study conducted through a geosurvey, a crowdsourcing tool targeted at a group of Kraków residents who experience mobility issues. The instrument allowed to obtain important data about travel behaviour and barriers to mobility, as well as to evaluate a number of proposed solutions aimed at improving travel conditions in the city. The data can be used in the process of planning sustainable mobility in the future and the unique value of the study resides in the fact that it addresses the transport needs of child caregivers, a subject which has not yet been thoroughly studied.

2 Origin, Objective, and Methodology

The geosurvey was prepared in the framework of a research project aiming to determine the needs and expectations of people with reduced mobility, especially child caregivers, with respect to the public transportation system in Kraków, and to use the data as a basis for developing relevant recommendations for city authorities. The project was inspired by "I Wózkowa Masa Krytyczna", a campaign organized on 15 October 2016 in Kraków by the MamaTon Klub Mam Foundation. In the framework of the initiative, people with reduced mobility marched through the streets of a Kraków neighbourhood in order to attract public attention to the stumbling blocks that parents with little children, seniors, and people with disabilities come up against in the area. Their insights provided a point of departure for an attempt to gather further information on the needs and issues experienced by sensitive groups and to study the scale of the problem in the city at large.

The geosurvey consisted of two sections. The first was a questionnaire, which focused on textual data on the behaviour, needs, and opinions of transport users. Since the research team was particularly interested in the child caregiver group, the questionnaire divided respondents into three user types: child caregivers, people in wheelchairs, and individuals with other mobility issues. The scope and focus of survey questions differed slightly for each group; the common issues concerned: their preferred means of transportation, reasons for the choice, perception of proposed solutions to improve travel conditions, and suggestions concerning possible mobility improvements. The questionnaire had been constructed in consultation with a representative of the MamaTon Klub Mam Foundation.

The other part of the geosurvey was an interactive map that allowed to gather spatial data about various barriers to mobility throughout the city. Respondents were asked to use the tool to identify trouble spots on the map and rate their inconvenience. The result was a database of spatial indicators subsequently analyzed with a broad range of GIS instruments. When mapping, respondents could choose from a list of predefined barriers to walking or cycling (e.g. lack of sidewalks, lack of bicycle paths), as well as define other problems and obstacles.

Tight cooperation with the Active Mobility Team at Kraków Municipal Infrastructure and Transport Board made it possible to launch the geosurvey on the official website of the Mobile Kraków project. A link to the survey was also posted on social media and relevant announcements were publicized in the media (on regional and national TV, online television, radio, and articles on info portals), as well as sent to Kraków nurseries and preschools.

3 Survey Findings

In the pilot phase (May–August 2017), during which the following results were obtained, the questionnaire was correctly filled out by 312 respondents. 73.1% were child caregivers (parents or designated custodians). Individuals in wheelchairs represented 5.1%, people with other physical disabilities (e.g. limb injuries, crutches, persistent pain) accounted for 6.4% and their caregivers for a further 4.2%; 3.2% were seniors. 8.0% of surveyed respondents reported other causes of reduced mobility (e.g. heavy luggage, pregnancy). The largest age group included individuals between 31 and 40 years of age (54.2%), followed by those in the 25–30 bracket (15.7%). The 17–20 age group was the least represented among respondents (1%). Women were more likely to take part in the study (60.6%). These sociodemographic characteristics can be traced back to the research team's focus on the needs of child caregivers and the promotion of the survey among the group. 63% of caregivers had children of up to 3 years of age, 21% – between 4 and 7, and the final 16% – over 7 years old.

3.1 Preferred Means of Transport for People with Reduced Mobility

One of the most important aspects studied by the survey had to do with the preferred means of transport for daily travel. Figure 1 presents the share of various options in the daily travel of three distinct groups: respondents at large, caregivers travelling with children, and, due to the similar nature of experienced difficulties – a collective group of seniors, individuals with physical disabilities, and their caregivers.

As shown in the figure, a majority of respondents opt for public transportation; this holds particularly true for seniors, individuals with physical disabilities, and their caregivers (60% of all trips in this group). The second most popular choice is a privately owned car, which is used with equal frequency as public transportation in the caregiver group. Approximately 1/5 of the time, individuals with reduced mobility decide to walk. 6% of all trips are done by bicycle, a means of transportation chosen mainly by child caregivers. 5% of seniors, individuals with reduced mobility, and their caregivers decide in favour of a taxi or a rented automobile. Differences could be observed between men and women; the latter were more likely to choose to walk or go by bus or tram, while men preferred privately owned cars and bicycles.

Respondents were also asked to name the reasons for their choice; they could indicate a maximum of three different motivations. The choice of a privately owned car was shown to largely depend on comfort: as many as 70.1% of respondents picked it as a reason. The short travel times and ease of use also proved essential, while caregivers tended to emphasize the importance of it being adapted to the needs of children. Taxis and rented cars were usually chosen for the sake of comfort and for lack of other options. Public transport, on the other hand, was preferred for a greater variety of reasons. 46.3% of all users mentioned its affordability and a further 45.5% emphasized ease of access. Other important reasons included the lack of other options, ease of use, and for child caregivers, the fact that the means of public transport is adapted to the needs of children. For bicycles, most respondents indicated the short travel time (57.9%), ease of use (52.6%), and affordability (47.4%). Walking is usually chosen because of the short distance to destination (57.1%), affordability (48.2%), and ease of use (33.9%).

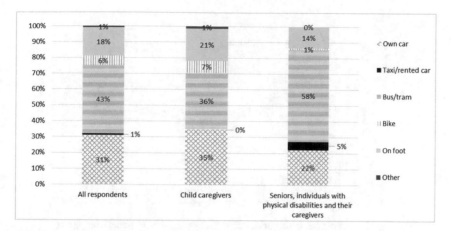

Fig. 1. Share of different means of transportation in the daily travel of people with reduced mobility.

3.2   Barriers to Mobility

The second part of the survey consisted of a map designed to collect data on specific barriers to mobility, whether by bike or on foot, and their location in the city. More than 1200 trouble spots were identified. Most barriers involved the lack of or discontinuous sidewalks (17.6%), closely followed by parked cars encroaching on the pedestrian walkway (15.3%). The lack of bike paths accounted for a further 12.7%. Based on collected data on barrier location, maps were drawn up to illustrate the areas in the city where they are particularly concentrated. These maps are shown in Figs. 2, 3, and 4.

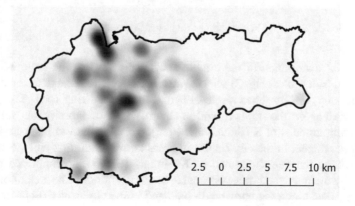

Fig. 2. Areas which lack or have discontinuous sidewalks.

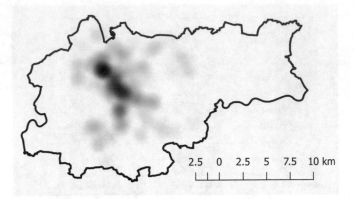

Fig. 3. Sidewalks with parked cars that encroach on pedestrian space.

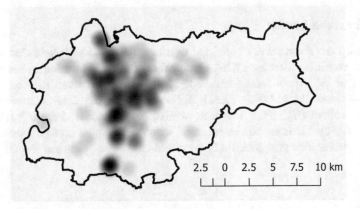

Fig. 4. Areas that lack bike paths.

The spatial distribution of the trouble spots indicates that the issue of missing or discontinuous sidewalks (Fig. 2) affects, to a lesser or greater extent, many different areas of the city, mostly those located beyond the second ring road. The problem is especially serious on the northern and southern flanks of the city, as well as in the central neighbourhoods of Kazimierz and Podgórze. Cars tend to limit the freedom of pedestrian movement in the city centre: in areas between the first and the second ring road, as well as the neighbourhood of Kazimierz (Fig. 3), as confirmed by surveys conducted by the municipal transport management institution. The lack of bike paths is particularly noted along the north-south axis and in the city centre (Fig. 4). More bike paths are also needed along the east-west axis.

Respondents also rated the inconvenience of each barrier on a scale, from 1 to 5 where 1 stood for a very low and 5 for a very high degree of inconvenience. Average ratings are shown in Table 1.

Table 1. Inconvenience of selected barriers indicated by survey respondents.

Barrier	Inconvenience score
Lack/failure of lifts or elevators	5.00
Inadequate lighting	5.00
Narrow pedestrian refuge islands	4.50
Dangerous actions of other transport users	4.09
Lack of/discontinuous sidewalks	3.90
Narrow sidewalks	3.76
Lack of ramps	3.75
Lack of pedestrian crossings	3.73
Uncomfortable ramps	3.57
Sidewalks blocked by parked cars	3.54
Reduced visibility during walking or cycling	3.51
Low sidewalk surface quality	3.48
Lack of bike paths	3.46
Obstacles along the route, e.g. road signs, garbage cans, etc.	3.33
High curbs	3.29
Traffic lights not adapted to the needs	3.00
Narrow bike paths	2.80

The most inconvenient barriers were those least frequently marked on the map: the lack or failure of lifts and elevators, narrow pedestrian refuge islands at pedestrian crossings, as well as inadequate lighting. Other major inconveniences included the dangerous actions of other transport users, such as, for instance, speed limit violations, especially in residential areas. Nearly all barrier types received a score of 3.0 or more, which means that survey respondents experience them as really inconvenient.

3.3 Significance of Solutions Aiming at the Improvement of Travel Conditions

Respondents were also asked to rate proposed solutions to improve travel conditions (other than by car) on a scale from 1 to 5, where 1 meant an insignificant and 5 a highly significant solution. Average ratings are shown in Table 2.

From the perspective of survey respondents with reduced mobility, the most significant solutions have to do with the ease of pedestrian travel: wide, well-kept sidewalks with comfortable surfaces, direct access to most destinations and public transport stops, as well as the presence of ramps. Other important aspects concern the conditions of travel on public transport, especially the availability of low-floor fleets with sufficiently wide doors, and, last but not least, comfortable waiting conditions. The ratings of all suggested solutions centre around the "rather significant" score. Some rating differences were observed for individuals in wheelchairs, who attached the greatest importance (an average score of more than 4.5) to low-floor fleets with wheelchair ramps, the good condition of sidewalks, their sufficient width, as well as the presence of ramps and lifts near underground crossings and footbridges. Issues not listed above, but

Table 2. Significance of solutions aimed to improve travel conditions (results for all respondents).

Solution	Degree of significance
Ensuring the good condition and even surface of sidewalks	4.51
Ensuring that sidewalks are sufficiently wide, without obstacles	4.51
Ensuring direct pedestrian access to basic travel destinations	4.45
Introducing a fully low-floor public transport fleet	4.41
Ensuring direct and convenient pedestrian access to public transport stops	4.36
Ensuring that the doors of public transport vehicles are sufficiently wide	4.35
Mounting ramps and adapted outdoor stairs	4.32
Introducing bus and tram stop shelters to protect against wind and rain	4.25
Using curbs that allow public transport vehicles to pull up to the edge of the platform	4.25
Ensuring that there is adequate space at the stop and the platform is wide enough	4.22
Separating pedestrian walkways from car traffic	4.12
Traffic calming measures	3.97
Presence of passenger lifts (elevators) by underground crossings and footbridges	3.94

frequently brought up by survey respondents, included the pressing need to develop the bike infrastructure and increase the frequency of trams and buses. Survey respondents also commented on the relative underdevelopment of the transportation network in the peripheries as compared with the city centre. Another aspect that requires improvement is the issue of mutual respect among public transport users.

4 Conclusions

Surveys that study the needs of individuals with reduced mobility usually focus on seniors and people with physical disabilities. In contrast, findings presented in this paper also include the group of child caregivers. Data for analysis was collected through an online crowdsourcing tool, which allowed to detect, map, and address major issues experienced by several sensitive social groups. The instrument was designed to examine actual needs and spur respondents to action, at the same time fostering a dialogue with local institutions in charge of public transport and sustainable mobility planning.

The survey demonstrated that Kraków residents with reduced mobility are most likely to opt for public transport (mainly due to its affordability and easy access) and privately owned cars (mostly because of comfort). Barriers to walking or cycling usually have to do with discontinuous sidewalks and their total absence, as well as obstacles such as parked cars and the lack of bike paths. Those rated as particularly inconvenient include lack or failure of ramps and lifts, inadequate lighting, and narrow pedestrian refuge islands on pedestrian crossings. In order to improve the mobility of sensitive groups, it is thus imperative to improve the conditions of pedestrian travel and public transportation by developing and enhancing infrastructure, as well as adapting transport services to their specific needs. Last but not least, it is no less important to develop local cycling infrastructure and take measures to build an atmosphere of mutual respect among the users of the public transportation system.

References

1. European Commission DG Energy and Transport: Action Plan on Urban Mobility, Brussels (2009)
2. Arsenio, E., Martens, K., Di Ciommo, F.: Sustainable urban mobility plans: bridging climate change and equity targets? Res. Transp. Econ. **55**, 30–39 (2016)
3. Wefering, F., Siegfried, R., Bührmann, S.: Guidelines: Developing and Implementing a Sustainable Urban Mobility Plan. Rupprecht Consult - Forschung und Beratung GmbH, Cologne (2014)
4. Sierpiński, G.: Travel behaviour and alternative modes of transportation. In: Mikulski, J. (ed.) Modern Transport Telematics. TST 2011. Communications in Computer and Information Science, vol. 239, pp. 86–93. Springer, Berlin (2011)
5. Lindenau, M., Böhler-Baedeker, S.: Citizen and stakeholder involvement: a precondition for sustainable urban mobility. Transp. Res. Procedia **4**, 347–360 (2014)
6. Renn, O.: Participatory process for designing environmental policies. Land Policy **23**, 34–43 (2006)
7. Lindenau, M., Böhler-Baedeker, S.: Participation actively engaging citizens and stakeholders in the development of sustainable urban mobility plans (2016). http://www.eltis.org/sites/default/files/sump-manual_participation_en.pdf
8. Gil, A., Calado, H., Costa, L.T., Bentz, J., Fonseca, C., Lobo, A., Vergilo, M., Benedicto, J.: A methodological proposal for the development of natura 2000 sites management plans. J. Coast. Res. SI **64**, 1326–1330 (2011)
9. Starzyńska, B., Kujawińska, A., Grabowska, M., Diakun, J., Więcek-Janka, E., Schnieder, L., Schlueter, N., Nicklas, J.-P.: Requirements elicitation of passengers with reduced mobility for the design of high quality, accessible and inclusive public transport services. Manag. Prod. Eng. Rev. **6**(3), 70–76 (2015)
10. Pashkevich, A., Puławska, S.: Accessibility of transport service for people with restricted mobility: needs analysis for a special assistance service in Poland based on the German experience. Logistyka **4**, 1453–1462 (2015)
11. Bühler, C., Heck, H., Sischka, D., Becker J.: BAIM–information for people with reduced mobility in the field of public transport. In: Miesenberger, K., Klaus, J., Zagler, W.L., Karshmer, A.I. (eds.) Computers Helping People with Special Needs. ICCHP 2006. LNCS, vol. 4061, pp. 322–328. Springer, Berlin (2006)
12. May, A., Parker, C.J., Taylor, N., Ross, T.: Evaluating a concept design of a crowd-sourced 'mashup' providing ease-of-access information for people with limited mobility. Transp. Res. Part C: Emerg. Technol. **49**, 103–113 (2014)
13. ISEMOA: D1.5 Final Publishable Report: Accessible and energy-efficient mobility for all! (2013). http://www.isemoa.eu/docs/8/ISEMOA_D1_5_final_publishable_report.pdf
14. Jankowska-Karpa, D.: Assessment of public space and public transport accessibility in cities, municipalities and regions across Europe. Logistyka **3**, 1907–1913 (2015)

15. Ferrari, L., Berlingerio, M., Calabrese, F., Reades, J.: Improving the accessibility of urban transportation networks for people with disabilities. Transp. Res. Part C: Emerg. Technol. **45**, 27–40 (2014)
16. Shrestha, B.P., Millonig, A., Hounsell, N.B., McDonald, M.: Review of public transport needs of older people in European context. J. Popul. Ageing **10**(4), 343–361 (2017)
17. Hammon, L., Hippner, H.: Crowdsourcing. Bus. Inf. Syst. Eng. **3**, 163–166 (2012)
18. Mirri, S., Prandi, C., Salomoni, P., Callegati, F., Campi, A.: On combining crowdsourcing, sensing and open data for an accessible smart city. In: Proceedings of the 8th International Conference on Next Generation Mobile Apps, Services and Technologies (NGMAST 2014), pp. 294–299. IEEE, Oxford (2014)
19. Mobasheri, A., Deister, J., Dieterich, H.: Wheelmap: the wheelchair accessibility crowdsourcing platform. J. Open Geospatial Data Softw. Stand. **2**(1), 1–7 (2017)

Movement Analysis of Inhabitants in the Upper Silesia Agglomeration (Poland)

Damian Lach[(✉)]

Faculty of Transport, Silesian University of Technology, Katowice, Poland
damian.lach@polsl.pl

Abstract. Mobility preferences of residents of a specific area have a significant impact on the shaping of transport systems on a microscopic and macroscopic scale. Preference data is used to improve existing systems or design new ones that have an impact on improving the living standards of residents. This condition is one of the transport postulates. In order to present the analysis data on mobility preferences of the inhabitants of the Upper Silesian agglomeration were used.

Keywords: Transport systems · Public transport · Surveys · Mobility behavior

1 Introduction

Transport systems of agglomerations and individual cities are increasingly burdened with what, among others testify to the results of multi-aspect research [1–3]. The congestion on the elements of transport networks has a negative impact on the economic and social development of the areas concerned. Urbanized areas are characterized by constant dynamics of changes in terms of urban planning and mobility of the society. An important element in the organization of transport systems is a proper analysis of communication preferences of residents and immediate response to changes. The challenge of transport policy is to change such preferences and the basis for undertaking such activities is the analysis of economic and social factors affecting the mobility of residents [4–6]. The article presents an analysis of mobility preferences of a selected group of inhabitants of the Upper Silesian agglomeration. The Upper Silesian Agglomeration is characterized by specific conditions describing the operation of transport systems. One of them is the separation of individual systems in ticket tariffs. An example of this are individual tariffs for regional railways and for bus, tram and trolleybus. The existence of several organizers of public mass transport and various sources of financing of given transport systems may have an impact on this state of affairs [7, 8]. Surveys were carried out at selected measurement points. Places where research was carried out were selected on the basis of appropriate criteria, including the number of inhabitants of a given city and the number of passengers using individual public transport stops. The research was carried out in March and April 2017. During the interview with respondents, information about the source and purpose of the ongoing journey, travel behaviors or the structure characterizing a given respondent was obtained. The aim of the analysis is to present the mobility of a selected group of

© Springer Nature Switzerland AG 2019
G. Sierpiński (Ed.): TSTP 2018, AISC 844, pp. 85–94, 2019.
https://doi.org/10.1007/978-3-319-99477-2_8

agglomeration residents and characterize the directions of development of the transport system of the area. By obtaining information on the intermodality of the journey, appropriate proposals for existing transport systems were drawn up.

2 Movement Analysis

2.1 Area of Analysis

The analyzed research area of Upper Silesian agglomeration is a specific region in a country consisting of many urban centers that form a whole. Depending on the sources, the number of cities included in the agglomeration varies significantly. In the case of data from the Central Statistical Office, the number of cities included in the agglomeration is 19, and the population is 2,106 million people [9]. In this way, the main area to be analyzed was determined. The choice of cities influenced the designation of places for surveys. The main purpose of the survey was to collect information about mobility preferences and transport behaviors of habits of the Upper Silesian agglomeration. According to [10, 11] surveys should be carried out at points characterized by large passenger flows. These types of points were chosen for own surveys. Cities in which research has been carried out have a population of over 100000. The research was carried out in the period of March and April 2017. Table 1 presents the list of places and dates in which surveys were carried out.

Table 1. List of measuring points with numbers of surveys.

City	Name of measuring point	Date of measurement	Number of surveys
Gliwice	Train Station (platforms)	21.03.2017 (from 7:00, morning rush hours)	42
	Piastów Square	21.03.2017 (from 13:00, afternoon rush hours)	30
Zabrze	Train Station	22.03.2017 (from 13:00, afternoon rush hours)	23
	Goethego (all platforms)	22.03.2017 (from 13:00, afternoon rush hours)	30
Ruda Śląska	Ruda Chebzie	24.03.2017 (from 7:00, morning rush hours)	21
	Chebzie Pętla	24.03.2017 (from 13:00, afternoon rush hours)	25
Bytom	Train Station	29.03.2017 (from 13:00, afternoon rush hours)	20
	Bus Station	29.03.2017 (from 13:00, afternoon rush hours)	40
	Sikorskiego Square	29.03.2017 (from 13:00, afternoon rush hours)	38

(*continued*)

Table 1. (*continued*)

City	Name of measuring point	Date of measurement	Number of surveys
Chorzów	Market Square	4.04.2017 (from 7:00, morning rush hours)	32
	Chorzów Batory (train, bus, tram)	4.04.2017 (from 13:00, afternoon rush hours)	20
Katowice	Market Square	31.03.2017 (from 7:00, morning rush hours)	42
	Korfantego Avenue	5.04.2017 (from 13:00, afternoon rush hours)	34
	Piotra Skargi Street	5.04.2017 (from 13:00, afternoon rush hours)	32
	Station (underground bus station)	7.04.2017 (from 7:00, morning rush hours)	33
	Station (train platforms)	7.04.2017 (from 13:00, afternoon rush hours)	33
	Wolności Square	12.04.2017 (from 13:00, afternoon rush hours)	34
Tychy	Public Transport Station	19.04.2017 (from 13:00, afternoon rush hours)	30
	Train Station (train platform)	21.04.2017 (from 7:00, morning rush hours)	20
Sosnowiec	Train Station (bus and tram platforms)	26.04.2017 (from 13:00, afternoon rush hours)	25
	Train Station (train platforms)	28.04.2017 (from 7:00, morning rush hours)	30
Dąbrowa Górnicza	Centrum	28.04.2017 (from 13:00, afternoon rush hours)	34

2.2 Description of the Survey Form

The survey was structured in such a way that it was properly readable by the respondent. It has been divided into 4 parts. In each part information was collected that affected the further course of the study. The first part of the survey concerns information about the place of residence and destination of the respondent's journey. Questions were asked to obtain information about:

- commune/city of residence (if commune/city is outside of the expected criterion, then filled out only part I and II);
- destination place.

The second part of the survey concerns the structure of the research sample. The information obtained concerns the gender, age, professional activity of the respondents and general mobility behavior of the respondents. The third part of the survey concerns detailed information about the source and destination of individual respondents. As part of this part, information was also collected on the motivation and course of the current journey. Detailed information applies:

- district or estate of a given respondent;
- public transport stop or station of the beggining of the journey;
- district or estate of destination place;
- public transport or station of the beggining of the return trip;
- travel motivation;
- the course of the current journey.

In addition, a question was asked about the frequency of using these means of transport: pedestrian, bike, car, bus, tram and train. In the fourth part of the survey, respondents were asked about opinions on the potential form of integration of public transport organizers and general communication preferences related to the possible existence of such a fact.

2.3   Survey Results

In the graphic form, the characteristics of the obtained information on mobility preferences of the inhabitants of the Upper Silesian agglomeration were drawn up. Figure 1 presents information about the place of residence of the respondents. 92% of respondents declared that they live in the Upper Silesian agglomeration.

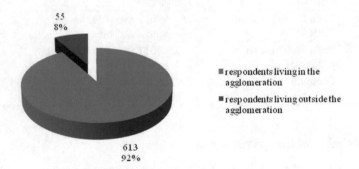

Fig. 1. Data of place of residence

Figure 2 presents data on the respondent's travel destinations. 93% of the persons questioned informed that the destination is located in the agglomeration. This part of respondents was included in further analysis.

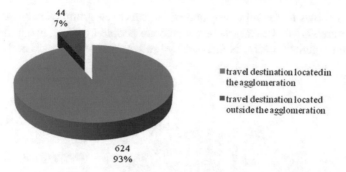

Fig. 2. Data of destination

Figure 3 presents the age structure of individual respondents. The largest group are people from 19–26 and it is 50% of all respondents.

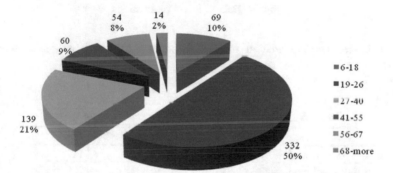

Fig. 3. Structure of respondent's age

Figure 4 shows the professional structure of respondents. The group of working people has the largest share in travels. Slightly less public transport services are used by students.

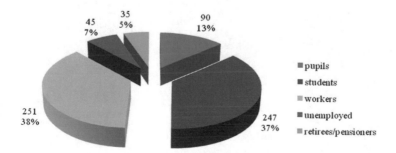

Fig. 4. Occupational structure of respondents

Figure 5 shows the travel motivation of the surveyed group of respondents. Travel motivations are varied. The largest percentage are people traveling in a home-study and study-home relationship. This is dictated by the fact that learners are at the same time working people.

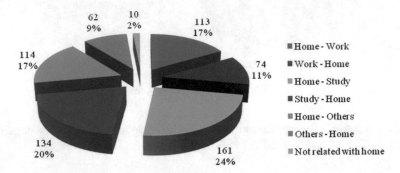

Fig. 5. Data of travel motivations

Table 2 presents information on the frequency of using particular means of transport.

Table 2. Frequency of using particular means of transport.

	Pedestrian	Bike	Car	Bus/trolleybus	Tram	Train
1–3 times a week	27	51	155	41	121	129
4 times and more	482	13	34	103	209	174

The information provided shows that respondents most often travel on foot. However, this may be due to the necessity to reach individual stops, stations or car parks. The data was shown on the Fig. 6.

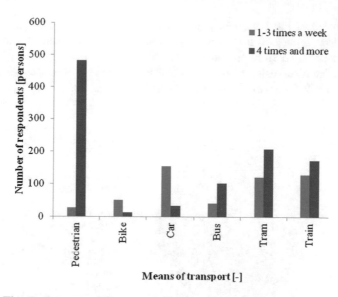

Fig. 6. Structure of frequency of using particular means of transport

Figure 7 shows the structure of travels made with the division into means of transport. Travel made with only one means of transport is a significant group of travels. The most trips were made by bus.

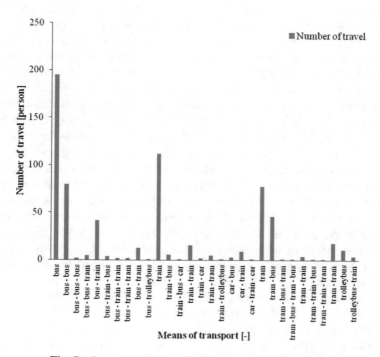

Fig. 7. Structure of travel divided into types of transport

On the basis of the presented graph, a pie chart was created (Fig. 8), which is presented below. It presents the percentage share of individual means of transport in the travel chains obtained from the survey. On its basis, it can be seen that the most movements were made using a bus. The value is as much as 49%. The least travel was done using a trolleybus and other means of transport (i.e. cars)

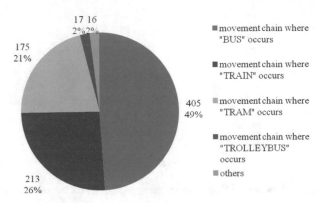

Fig. 8. Percentage of means of transport on trips

2.4   Analysis of Survey Results

The main objective of the analysis is to indicate factors that affect the mobility preferences of the selected area. Based on the obtained results, the structures of travels are depicted. The data show that the largest group of people doing travel are students of high schools and working people. In the case of students, this is due to the fact that there are concessions allowing travels to be made 50% cheaper in relation to the base price. In the case of employees, this is most likely due to the lack of an individual means of transport and/or the lack of convenient public transport connections for a selected group. This relationship is confirmed by the results regarding the age structure of the respondents. 332 respondents, or 50% of the whole surveyed group, are aged between 19 and 26. The analysis includes only trips that have their purpose in the Upper Silesian agglomeration. This means that people who travel outside of the agglomeration are not included. Such journeys were considered incidental travels. In the case of travel motivation, the largest part are those traveling in the relation home-study and study-home. Only 28% of trips were made to and from workplaces, which means that working people are also students. The largest part of the journey are journeys made inside the administrative borders of cities where survey research was carried out. Most of such trips have been made in the city of Katowice. It also results from the fact that in the given city there were the most measuring points. This is related to the character of the city, which is the main urban center in the Upper Silesian agglomeration. Such dependence was confirmed by the results regarding the number of trips made to Katowice from other cities. The most numerous group are journeys made from Gliwice and their number is 35. In other cases, the number of trips to Katowice is at a similar level. In the case of the city of Jaworzno, the number of trips is 0, because no research was carried out in this city. This is due to the criteria for selecting measurement points. Data regarding the number of trips in a given relation refer only to return trips from Katowice. The structure of the journey, devided into means of transport, shows that the largest group consists of journeys made with one means of transport. Most of the respondents used only the bus. The results show that the services offered by the organizers do not encourage intermodal travel. Combined journeys were made only in relationships where there are no direct connections.

3 Conclusion

The following conclusions were obtained from the analysis:

- questionnaire surveys have been carried out at the appropriate measurement points specified in the relevant criteria;
- measuring points at stops and stations referred to as a interchange node in the analysis have been compressed to one reference point (for example: Gliwice Plac Piastów and Gliwice Dworzec Kolejowy was defined in the analysis as);
- there are three organizers of public mass transport in the research area (KZK GOP, MZK Tychy, Koleje Śląskie);
- the most numerous group of respondents are working people and students;
- the largest number of people using public transport services is between 19 and 26 age;
- the analysis includes only trips with a purpose in the Upper Silesian agglomeration;
- travel motivations are varied, but the largest group are journeys made in the relation home-study and study-home;
- the majority of respondents made a trip in the Gliwice - Katowice relationship;
- the most trips were made using one means of transport. Combined journeys make up 40% of all registered journeys, most of which are journeys made with one transfer;
- over 31% of travels have been completed on foot;
- proper tariff and organizational integration will allow to increase the attractiveness of public public transport;
- integration will change the mobility behavior of the residents and encourage them to make combined journeys, which will have a positive effect on the entire agglomeration's transport system in terms of organization and economy;
- proper organization of public transport within the agglomeration will allow for more efficient use of vehicles and available infrastructure.

References

1. Macioszek, E., Lach, D.: Analysis of the results of general traffic measurements in the west pomeranian voivodeship from 2005 to 2015. Sci. J. Silesian Univ. Technol. Ser. Transp. **97**, 93–104 (2017)
2. Macioszek, E., Czerniakowski, M.: Road traffic safety-related changes introduced on T. Kościuszki and Królowej Jadwigi Streets in Dąbrowa Górnicza between 2006 and 2015. Sci. J. Silesian Univ. Technol. Ser. Transp. **96**, 95–104 (2017)
3. Macioszek, E.: The comparison of models for follow-up headway at roundabouts. In: Macioszek, E., Sierpiński, G. (eds.) Recent Advances in Traffic Engineering for Transport Networks and Systems. LNNS, vol. 21, pp. 16–26. Springer, Cham (2018)
4. Zawieska, J.: Inhabitants of Warsaw mobility preferences and behavior in the context of socio-economic transformations between 1993–2015. Transp. Miej. Reg. Ser. **3**, 17–23 (2017)
5. Daniels, P., Warnes, A.: Movement in Cities. Routledge, England (2007)
6. Paulley, N., Balcombe, R., Mackett, R., Titheridge, H., Preston, J., Wardman, M., Shires, J., White, P.: The demand for public transport: the effects of fares, quality of service, income and car ownership. Transp. Policy **13**, 295–306 (2006). Elsevier, Holland
7. Janecki, R., Krawiec, S., Sierpiński, G.: Publiczny transport zbiorowy jako kluczowy element zrównoważonego systemu transportowego Górnośląsko-Zagłębiowskiej Metropolii Silesia. Urząd Miasta Katowice. Series I, pp. 105–132 (2010)
8. Pucher, J., Kurth, S.: Verkehrsverbund: the success of regional public transport in Germany, Austria and Switzerland. Transp. Policy **2**, 279–291 (1995). Elsevier, Holland
9. Upper Silesian Conurbation According to the Central Statistical Office (2012)
10. Wyszomirski, O., Gromadzki, M.: Regular monitoring of public transport passengers' behavior - Gdynia case study. In: Conference "Intelligent public transport", Warsaw (2015)
11. Sierpiński, G., Celiński, I., Staniek, M.: Rejestracja zachowań komunikacyjnych w czasie rzeczywistym jako wsparcie organizacji i zarządzania transportem. Logistyka **3**, 5273–5280 (2015)

Impact of Cyclist Facility Availability at Work on the Number of Bike Commuters

Romanika Okraszewska[✉]

Department of Highway and Transportation Engineering,
Gdańsk University of Technology, Gdańsk, Poland
romanika.okraszewska@pg.edu.pl

Abstract. The article describes the results of research designed to establish whether cycle provision can influence the number of employees commuting by bike. To that end, employee surveys were conducted in three IT companies in 2012 and 2016. The questionnaire asked about travel behaviour and what the staff thought about their company's provision of cyclist facilities. Since 2012 each of the companies has moved its head office and the facilities for cyclists have improved. The studies from 2012 indicated a greater impact of an unseen level of organizational culture than that of artifacts such as facilities for cyclists at the company or its location relative to the system of city cycle paths. Results from 2016 indicate that when significant changes are made on the artifacts level the influence on the number of cyclists can be noticeable.

Keywords: Cycling commuter · Modal split · Organizational culture

1 Introduction

The cities of today are facing the challenge of meeting the growing mobility needs of their residents and reducing the number and intensity of conflicts that are caused by transport as it operates and continues to grow. Some of the main urban transport challenges identified by the European Commission include the need for a change of transport behaviour and how transport is viewed by city dwellers [1]. The emphasis should be on sustainable transport attitudes to be achieved by promoting walking, cycling and using public transport [2].

Before people's transport behaviour can be optimised, conditions must be in place to stimulate sustainable choices and decisions [3]. The entities that are best positioned for fostering a new urban mobility culture must be able to carry the message to large groups of society or large groups of people. These will, in particular, include the authorities (national and local) but also centres of education, religious groups, the media, on-line content creators and employers [4].

This article focusses on the role of employers in promoting their staff's transport behaviour. Those economically active represent an important share in the overall population of a city. Work commutes tend to be regular and are major contributors to traffic peaks. This is why a change in the behaviour of a group of commuters is likely to leave a mark on the city's traffic.

© Springer Nature Switzerland AG 2019
G. Sierpiński (Ed.): TSTP 2018, AISC 844, pp. 95–105, 2019.
https://doi.org/10.1007/978-3-319-99477-2_9

Efforts to promote sustainable transport and manage the demand for car trips for work commutes are usually delivered via mobility plans. This involves changing staff behaviour and attitudes and can be implemented at workplace level.

In literal sense, the main goal of mobility plans is to facilitate a modal shift toward more sustainable transport modes. Usually, the potential targets for the modal shift are case-specific and may depend on economic, spatial and infrastructural conditions [5]. Many of the examples described in the literature refer to public facilities/institutions [6] such as schools [7], universities [8], municipal bodies, hospitals [9], libraries, shopping malls [10] etc. as well as businesses located in urban areas [10]. The described effects vary from case to case and range from significant to negligible increases in pedestrian, bicycle or public transport share in all trips.

While mobility plans have been widely used worldwide for several years, mobility plans are quite scarce in Poland. Some companies, as part of their brand or efforts to attract employees, take steps to improve their cycle provision and are able to increase the number of their staff commuting by bike.

The article describes a case of three IT companies. Known for competing for talent, these companies have changed their cycle provision. The article describes the scope of change and how it influences the transport choices of the employees.

2   Previous Research and Conclusions

In 2012 three Gdańsk-based IT companies completed a survey designed to establish the number of people who cycle to work and the reasons behind their choice of transport mode. The main criterion for selecting the companies for the survey was to ensure uniformity of factors that are critical for transport choices staff make. The selection was based on the industry, age structure, size of company, location and cycling facilities in the workplace. The respondent companies were head offices of IT firms, located in the city centre, with good access to a network of cycle paths and a similar age structure of the staff. The differentiating feature between the firms was the level of cycle provision.

When analysing the factors that influence staff transport behaviour, reference was made to E. Schein's model of organisational culture. When viewed by an outsider, the model has three levels: artifacts, values and basic assumptions. The details of the method, results and conclusions from the research are described in article [11]. Studies helped to identify the relation between employer's organisational culture and staff transport choices. While the results of the study showed that organisational culture has a major influence, the expected and direct correlation between cyclist facilities and the number of bike commuters could not be established. When the values at level one and two of Schein's organisational culture are different, values have a stronger influence than artifacts, a finding established in the research.

In the last three years all of the companies surveyed in 2012 have moved to a new location. Getting to work has changed as well and the companies decided to improve their cycle provision as part of their branding efforts.

The repeat round of the survey was designed to understand how the corporate changes (new image and location) had changed the number of employees cycling to work. It was equally important to define the weight of how particular factors influenced staff transport behaviour. The data were used to evaluate the effectiveness of the measures based on the new modal split.

3 New Conditions

The main rationale behind the repeat commuter cycling survey was that the companies had moved to a new location, which is a significant employer-dependent factor that influences people's decision to cycle to work (Table 1).

Table 1. Factors that influence cycle commuting [5]

EMPLOYER DEPENDENT	EMPLOYER INDEPENDENT
Organisational culture:	
artifacts: workplace infrastructure meeting cyclist needs (cycle parking, lockers, shower)	Weather conditions
values: promoting environmentally-friendly transport behaviour, competing internally	Activities outside work (shopping, pre-school or school run, other activities and duties)
basic assumptions: respect for the natural environment, management setting an example	
Management style – staff mobility policy, staff trip plans, a system of cycle incentives	Location of employees' homes – distance to work, access to cycle networks.
Company location – distance to cycle network, access to public transport.	City cycling infrastructure.

The decision to move was largely prompted by the growing need for more space as the offices grew. Because office space in Gdańsk tends to concentrate in the Main Service Zone, new offices of all of the three companies are now in the city centre along Gdańsk's main road axis and close to the main cycle routes.

Analysis of the commute matrix, especially in the case of company C, shows that the new location is closer to the matrix's centre of gravity which should be reflected in the new average distance between home and work and reduce standard deviation.

As part of their efforts to attract employees and build their brand all the new offices come with lockers with sanitary and shower facilities and dedicated cycle spaces in the car park (level of artifacts). The new addresses meant a change not just for cyclists, because the location of the other transport sub-systems changed as well: SKM trains, trams, buses and available parking.

Conducted in company B in 2012 the interviews established that the level of values and basic assumptions suggests a pro-motorisation culture. When repeated in 2016, the interviews produced a different result. This may be the result of market pressure to change the image and demands from staff who prefer a healthy lifestyle. Many of the staff participate in cycle activities both those organised by the company and the city.

4 Objective and Scope of Research

Based on the 2012 surveys and three level division of E. Schein's organisational culture, three different company characteristics were obtained Fig. 1.

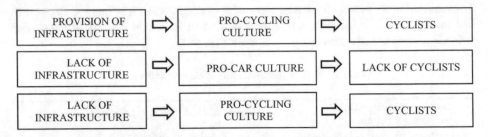

Fig. 1. Company characteristics identified in the 2012 survey based on the three level division of E. Schein's organisational culture.

With the changes of location and organisation, the companies' organisational cultures have unified both at the artifact level and the deeper levels of values and basic assumptions. The survey was repeated primarily to answer the question of whether the changes have changed the number of people cycling to work Fig. 2.

Fig. 2. The main research objective is to answer the question of how the companies' new organisational cultures will impact the number of bike commuters

The repeat survey after the companies had relocated was designed to:

- determine the frequency of staff cycling to work,
- determine the potential changes in the frequency of cycling to work as a result of the new location and new cyclist facilities,
- identify the significance of the employer-dependent factors for the number of bike commuters,
- identify the current modal split for the staff of the particular companies,
- compare staff perception of the changes and their effects with real data.

5 Method

To collect information about staff transport behaviour, the diagnostic poll was used conducted as an on-line survey. Surveys were sent to the staff of the same companies that participated in the 2012 survey. Google Forms was used to develop the

Table 2. Basic survey information for selected entities.

Firm	Population		Survey duration		Sample		Share of population	
	2012	2016	2012	2016	2012	2016	2012	2016
A	150	200	26.06–04.07	22–27.06	98	97	65%	49%
B	80	90	20.06–29.06	22.06–03.07	48	67	60%	74%
C	240	550	10.09–26.09	22–29.07	75	21	31%	4%

questionnaire and collect data. Respondents were invited to complete the questionnaire via the companies' internal e-mail addresses.

Table 1 gives basic information about the 2012 and 2016 surveys: date of survey publication of the different groups, survey duration and number of respondents, sample size following response verification and its percentage share in the group (Table 2).

The companies grew and took on more staff in the space of two years. Compared to 2012 the number of responses to the questionnaires, there were comparable in company A, almost 40% higher in company B and fewer in company C. In the case of companies A and B the sample can be considered representative. In the case of company C, however, the conclusions from the responses are likely to be unreliable and contain error.

6 Results

The data collected from 2012 and 2016 surveys were compared. The comparative analysis did not identify any significant changes in any of the firms as regards **age structure and gender**. As regards the **declared ability to cycle to work** there were no significant changes in firm A. The other two firms, however, reported a high number of people saying they cannot cycle to work. In firm B the declared inability to commute by bike changed from 0% to 25% and from 8% to 50% in firm C. This may have to do with more staff coming from outside the Tri-City who do not have a bike in their current home and no city bike to rent from the city of Gdańsk.

As regards the declared **distance to work** there were no significant changes in firms A or B. In firm C the average work-home distance changed (Fig. 3). The reasons for this include possibly the change of company address relative to the trip matrix's centre of gravity, employee turnover (50% of respondents did not work for the company in 2012) or an error due to the small size of the statistical sample.

Journey time to work is directly related to the distance and choice of transport mode (the percentages are the average values for the staff of all three companies):

- up to 10 min – journey time for about 5% of staff who live within 5 km and commute by bike (2%), car (2%) or on foot (1%),
- 10–20 min - journey time for about 25% of staff who live within 5 km and commute mainly by bike (11%) and on foot (2%), by car (2%) and by PT (3%) and by staff who live within 5–10 km who commute by car,

- 20–30 min - journey time for some people who live within 5 km and commute by car (2%), SKM train (1%), bike (1%) on foot (1%) and people who live within 5–10 km who commute by car (12%), bike (6%), public transport (2%), SKM train (1%) and those who live within 10–15 km and more and commute mainly by car (10%), SKM train (5%), public transport (1%) and by bike (1%),
- 30–60 min – journey time for people who live within 5–10 km from the office, commute by car (4%), SKM train (1%), PT (1%), people who live within 10–15 km from the office, who commute by car (2%), by bike (1%), public transport (1%), people who live more than 15 km from the office who commute by car (8%), SKM train (5%) or bike (1%),
- more than an hour – journey time for people who live more than 15 km from the office and commute by car (2%) or SKM train (2%).

Figure 4 shows the differences between the companies in how they declared journey times.

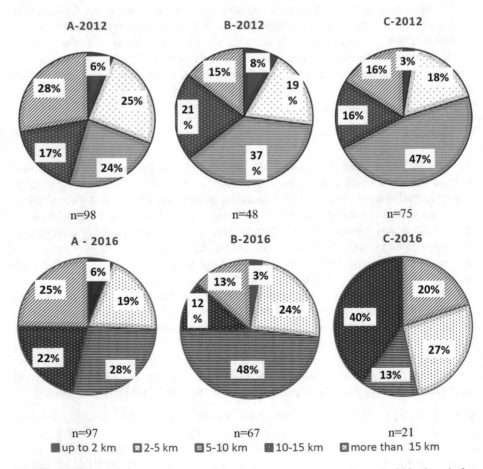

Fig. 3. Comparison of the home-work distance between the companies and within them before and after the move

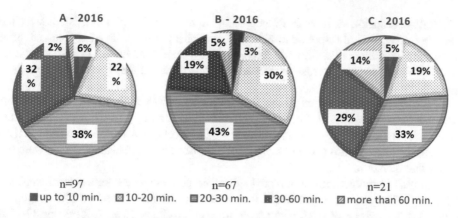

n=97   n=67   n=21

■ up to 10 min.   ▨ 10-20 min.   ▤ 20-30 min.   ▩ 30-60 min.   ▨ more than 60 min.

Fig. 4. Comparison of journey times after the companies have moved office.

Changes in the **frequency of cycling** to work. There were no significant changes in firm A. In firm B there has been a significant increase in people cycling to work all year round (from 0% to13%) with a lower share of people not cycling at all (from 92% to 33%). In firm C the change is significant with more people not cycling to work at all (from 39% to 71%), significantly fewer people cycling sporadically or frequently (from 58% to 24%), the share of people for whom using a bicycle every day is a habit remained without major changes (from 4% to 5%) Fig. 5.

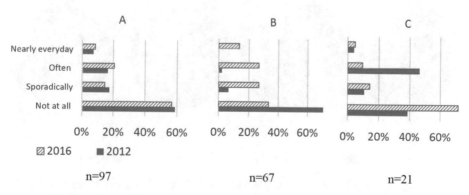

n=97   n=67   n=21

Fig. 5. Changes in the frequency of cycling to work in firms A, B and C.

Respondents' own assessment of the changed frequency of cycling confirms the trends observed. In firm A 69% of respondents said that the new location did not change how often they cycle with 13% of people saying that they cycle more often and 19% saying they cycle less often. In firm B 57% of respondents declare the frequency has not changed, 3% say the frequency has dropped and as much as 40% of respondents claim they cycle to work more often after the company has moved office. Firm C respondents were not asked that question.

Staff perception of how many people cycle to work is quite consistent with the data. In firm A 52% of respondents believe that cyclist numbers have not changed, 36% think they have increased with 12% claiming they have gone down. In firm B 93% of staff have recorded an increase in cyclists. No one believed that cyclist numbers have gone down.

Factors that have an effect on the frequency of cycling to work. In firm A respondents reported the new distance as the main determining factor with respondents in firm B suggesting that cyclist facilities in the workplace were important such as lockers, shower facilities and a safe place to park the bike. Firm C respondents were not asked that question.

Analysis of **modal split**. The modal split is based on answers for a typical weekday (Tuesday). The survey asked about journeys to work last Tuesday. The responses showed a high share of cycling for trips to/from work, A-19%, B-23% and C-15% respectively. When asked about the frequency of cycling to work the responses suggested that the bike is a popular mode of transport. In 2016 in the firms A-8%, B-13% and C-5% of respondents identified themselves as everyday bike users (Fig. 5). To recap, cycling accounts for a high number of staff journeys at about 15–23% and is clearly more prevalent than for the overall population of Gdańsk (6%) [12] (Fig. 6).

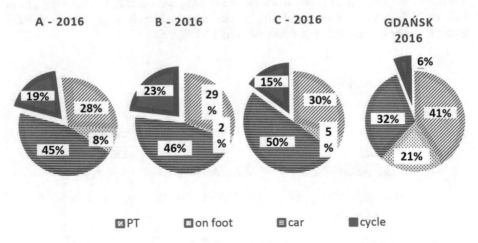

Fig. 6. Comparison of modal split in firms A, B and C and whole city [13]

The **reasons** why cycling is used for journeys mainly include fun, daily portion of exercise, directness, costs, environmental considerations, flexibility and convenience. When arguing the case for car journeys, directness, travel time and flexibility are most often quoted. When asked why they do not cycle, respondents mostly suggested having to run errands on the way to/from work, having to get extra ready before and after cycling, extra baggage being a problem, the weather and the effort it takes due to the distance, topography or fitness.

Changes as a Result of the New Locations The survey asked firms A and B respondents to explain how certain conditions changed as a result of the move. These include: car parking, commuting by car, using public transport and cyclist facilities.

Firm A respondents believe car parking has improved with driving to work becoming more difficult. Staff responses suggest that the new location is poorly served by public transport. While respondents have noted improved availability of lockers, shower facilities and parking spaces, they were not happy with the security of their bikes and considered it less safe compared to the previous office.

Firm B respondents said that all the conditions have improved across the board. On a scale of −3 to +3 their opinion on the availability of lockers and shower facilities was a unanimous +3.

To assess whether staff behaviour can be influenced, firms A and B respondents were asked to say how **ready they are to use the car less** ad instead switch to solutions such as carpooling. In both firms the responses were similar. About 30–40% of staff do not feel they should use the car less at all. Another 20–30% would like to do that but do not think it possible. While others say yes, their final decision is based on other conditions. As regards carpooling on average 40% of respondents said no with 40% quite likely to say yes and 20% having no opinion. There were no major differences between participating in carpooling as a driver or passenger.

7 Conclusion and Discussion

Because the offices are in similar locations, the external conditions for commuting to work have uniformed as have cyclist facilities (level of artifacts – cycle parking, lockers, shower facilities). With consistent organisational cultures across all levels the three companies have a fairly similar and high share of cycling on the way to and from work.

Based on the 2012 survey it was found that when Schein's level one and two of organisational culture feature different values, values tend to have a stronger influence as opposed to artifacts. Repeated in 2016, the survey showed that, as exemplified by firm B, when artifacts change and the organisational culture is communicated consistently, they have a significant effect on staff transport behaviour.

The survey showed that in all the three companies the popularity of cycling as a means of transport has grown. The decision to commute by bike is clearly dictated by workplace location and the provision of cyclist facilities. As the location changes, so does the distance from home to work. Analysis of the relation between the choice of mode of transport and the distance shows that:

- short trips up to 2 km, mostly on foot or by bike,
- short trips between 2–5 km, about 50% are by bike,
- cycling is the most popular mode for trips between 2–5 km,
- long trips above 15 km are mainly by car or PT (SKM trains in Gdańsk),
- long trips are hardly ever by bike.

The significance of "distance" for people's individual choices of mode of transport suggests that this factor should definitely be considered when companies look for a location. The decision should also be related to the origin-destination matrix.

When the car is the preferred mode the decision is determined mainly by factors such as directness and travel time. As we know from respondents, the differences between travel time by car and by bike on the same distances are small and the biggest on distances between 5–10 km, which is when the car is seen as faster by on average 5–10 min. This is why as well as providing cyclist facilities, employers should do more to promote their staff's sustainable transport behaviour and change how they see the bike as a means of transport. Staff should be offered some incentives to try and perhaps get used to this mode of transport.

From a comparison of how the companies' characteristics has changed at all of E. Schein's levels of organisational culture and the resulting modal split, we can see that having a consistent organisational culture across all levels may stimulate sustainable transport behaviour.

To obtain more detail on the significance of company location for transport behaviour similar research will be conducted in an IT company, located on the outskirts of the city and a high degree of cyclist facility provision.

The potential which comes from staff mobility management should be addressed by sustainable urban mobility policy (SUMP) as regards possible incentives for employers to help form sustainable transport behaviour of staff and implement mobility plans.

References

1. European Commission: Green Paper. Towards a new culture for urban mobility, no. 9. Brussels (2007)
2. European Commission: Roadmap to a Single European Transport Area - Towards a competitive and resource efficient transport system, Brussels (2011)
3. Okraszewska, R., Romanowska, A., Wołek, M., Oskarbski, J., Birr, K., Jamroz, K.: Integration of a multilevel transport system model into sustainable urban mobility planning. Sustainability 10(2), 479 (2018)
4. Okraszewska, R., Nosal, K., Sierpiński, G.: The role of the polish universities in shaping a new mobility culture - assumptions, conditions, experience. Case Study of Gdansk University of Technology, Cracow University of Technology and Silesian University of Technology. In: ICERI2014 Proceedings, 7th International Conference of Education, Research and Innovation, pp. 2971–2979 (2014)
5. Norfolk Country Council: Travel Plan targets for modal shift as at December 2015 (2015)
6. Making travel plans work. Lessons from UK case studies, London (2002)
7. Peddie, B., Somerville, C.: Travel Behaviour Change through School Travel Planning: Mode Shift and Community Engagement – Results from 33 Schools in Victoria, Paper presented to 28-th Australasian Transport Research Forum, Sydney, Australia (2005)
8. Dell'Olio, L., Bordagaray, M., Barreda, R., Ibeas, A.: A methodology to promote sustainable mobility in college campuses. Transp. Res. Procedia 3, 838–847 (2014)
9. Petrunoff, N., Rissel, C., Wen, L.M., Xu, H., Meikeljohn, D., Schembri, A.: Developing a hospital travel plan: process and baseline findings from a western Sydney hospital. Aust. Heal. Rev. 37(5), 579 (2013)

10. Petrunoff, N., Wen, L.M., Rissel, C.: Effects of a workplace travel plan intervention encouraging active travel to work: outcomes from a three-year time-series study. Public Health **135**, 38–47 (2016)
11. Okraszewska, R.: The impact of organizational culture on employees' decisions on the selection of bicycles as a form of transportation to work, in the context of creating a new culture for urban mobility, based on an example of three IT companies located in Gdansk. Logistyka **4**, 3101–3110 (2014)
12. Okraszewska, R., Birr, K., Gumińska, L., Michalski, L.: Growing role of walking and cycling and the associated risks, MATEC Web Conference, vol. 122, p. 1006 (2017)
13. Group work: Report on the Comprehensive Traffic Research in Gdansk (in Polish), Sopot-Warszawa (2009)

Analysis of Factors Affecting Non-compliance with the Red Light Signal at City Intersections Equipped with Traffic Signaling

Tomasz Szczuraszek and Radosław Klusek[✉]

Faculty of Construction, Architecture and Environmental Engineering,
University of Technology and Life Sciences, Bydgoszcz, Poland
{zikwb,radoslaw.klusek}@utp.edu.pl

Abstract. The main causes of road collisions in Polish cities include disrespecting the priority of way rule and failing to keep safe distance between vehicles, however, according to statistics, running a red signal is another common cause of road accidents (the cause of 3.9% accidents and 1.6% collisions). The authors of this study have made an attempt to identify the factors which condition this kind of behavior and the degree of non-compliance with the traffic signaling by traffic users, including drivers, cyclists and pedestrians. Research was conducted for each group of the road users. Investigations of incidents connected with going/driving through a red signal were conducted. The total number of road users of the analyzed traffic stream, the number of persons who failed to comply with traffic signaling, their gender and age, traffic conditions as well as time that has passed since traffic signaling turned on until running a red signal. The tests results show that it is the amount the time the motorist have to spend waiting to leave an intersection has the largest influence on non-compliance with traffic signaling.

Keywords: Traffic signaling · Road traffic safety · Red signal

1 Introduction

Although, since 1997 the number of accidents in Poland has been systematically decreasing, Poland is still at the top of European Union countries where the risk of losing life in a road accident in the highest [1]. According to statistics [1], as many as 70.9% of road accidents in Poland are reported to have happened in a built-up area. The main causes of these events are non-compliance with priority of way and failing to keep safe distance between cars. However, according to statistics, running a red signal is another common cause of accidents, accounting for 3.9% of accidents and 1.6% of collisions [2]. Failing to obey traffic signaling or its equivalents can be reported for all groups of road users [3]. Lots of research concerning non-compliance with the red signal has been done worldwide [3–7]. The research results provide the basis to draw a conclusion that the level of non-compliance with traffic signaling significantly differs from country to country. This is attributed to different behaviors of road users in different countries and different approach of particular societies to respecting the law.

© Springer Nature Switzerland AG 2019
G. Sierpiński (Ed.): TSTP 2018, AISC 844, pp. 106–115, 2019.
https://doi.org/10.1007/978-3-319-99477-2_10

The authors have made an attempt to identify the conditioning and level of non-compliance with the red signal by motorists, cyclists and pedestrians in Poland. Gender and age have been analyzed as factors affecting the traffic users' failure to follow traffic signaling. Moreover, the traffic volumes and the number of cars waiting to enter the intersection (which had an impact on the time of waiting for entering the intersection) were additionally analyzed as well as the type of vehicles used by the drivers. Behavior of cyclists, however, was analyzed in terms of the length of the red signal illumination, whereas, the length of crosswalk was taken into consideration in the case of pedestrians. All investigations were performed in Bydgoszcz, a city with population of app. 350 thousand people.

2   Non-compliance with the Red Signal by Car Drivers

Non-compliance with the red signal for car drivers was performed at six intersections located in the center of Bydgoszcz. Observance ware carried out on channelized intersection and intersection with a central island with different numbers of lanes on the inlets. The major goal of the research was to check the number of vehicles running a red at a selected intersection, depending on the length of queue a given vehicle had to wait in to enter the intersection. These two parameters have a large influence on the time a driver needs to wait to cross the intersection. The traffic volumes at the intersection inlets with respect to their structure and the length of car queues standing before the traffic signaling was measured at each intersection. Moreover, the drivers' gender and age were recorded. The investigations were carried out on Tuesdays, Wednesdays, Thursdays and Sundays. Sunday measurements were performed in order to distinguish the traffic volumes level and the length of car queues on the intersection inlets. Measurements were performed the morning and afternoon rush hours, (6.30–8.30, 15.00–17.00) and in between the rush hours (15.00–17.00).

The results of research on the impact of traffic volumes at the intersection inlets on the number of drivers who disrespected the red signal are shown in Fig. 1. They indicate that the number of motorists running a red signal is growing along with the rise of traffic volumes at the intersection inlets. Such a situation was reported for all the analyzed intersections. According to the authors, the growth in the number of motorists running red signals for higher traffic volumes is caused by the prospect of wasting more time by drivers which irritates them and makes them upset. Under such circumstances, drivers tend to act more hastily to cross the intersection as quickly as possible. This thesis is also confirmed by the results of research on the influence of the length of car queue at the intersection inlets, which are presented in Fig. 2.

Also the types of vehicles used by drivers who failed to respect the red signal were analyzed. These results are presented in Fig. 3. Motorcyclists were found to be the group of road users who most frequently failed to respect the red signal, they accounted for 3.9% of all red signal offenders. The significant share of this group in traffic rules offenders is connected with high dynamics of the vehicles they use.

In this case the causes are to be looked for in travel destinations of these drivers. They often travel on business and have time rigors to be observed. Therefore, standing and waiting for traffic signaling to change makes them nervous and inclined to take up

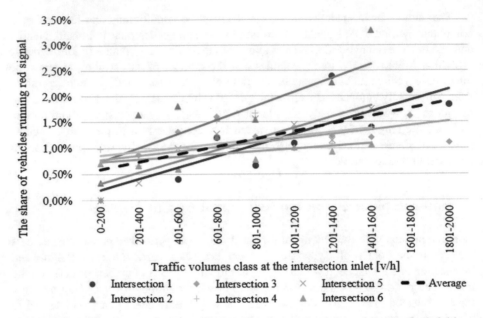

Fig. 1. Dependence between traffic volumes at the intersection inlet and the number of drivers running red signal

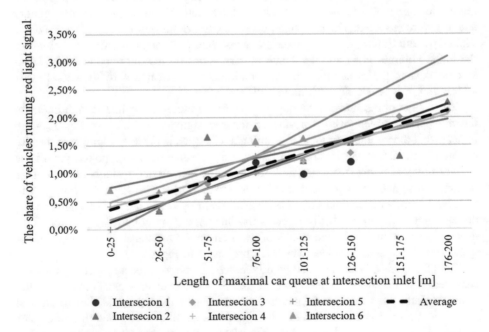

Fig. 2. Dependence between the maximal car queue at the inlet and the number of drivers running red signal

more risk while crossing an intersection [8]. No red violation was reported for drivers of low speed vehicles (low movement dynamics).

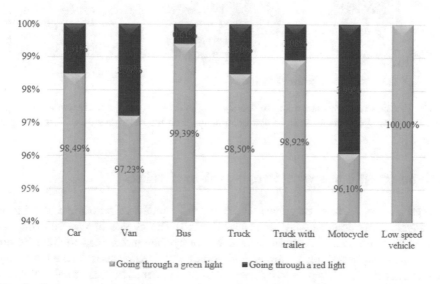

Fig. 3. Percentage share of drivers running a red signal in dependence on the vehicle type

Other factors contributing to non-compliance with traffic signaling are gender and age of road users, therefore, the authors also took into consideration these factors, though they were determined according to subjective assessment of the observer. The results of the research on the gender and age of road users disrespecting the red signaling are shown in Figs. 4 and 5. The results show that nearly 6% of man fail to observe the traffic signaling at intersections, whereas, there are half fewer women who behave in this way. The drivers aged 35–55 (6.90%) were found be the most frequently disrespecting the traffic signaling, whereas drives of age group under 35 (3.1%).

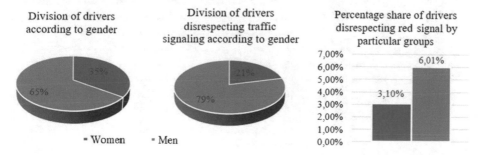

Fig. 4. Division of drivers according to gender/division of drivers disrespecting red signal according to gender/percentage share of disrespecting red signal by drivers in a given group of road users

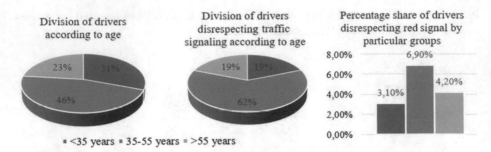

Division of drivers according to age

Division of drivers disrespecting traffic signaling according to age

Percentage share of drivers disrespecting red signal by particular groups

■ <35 years ■ 35-55 years ■ >55 years

Fig. 5. Division of drivers according to age/division of drivers who fail to respect the red signal according to age/percentage share of non-compliance with the red signal by drivers of a given group

3 Non-compliance with Red Signal by Cyclists

The research on cyclists' compliance with the red signal was performed at 6 bicycle crossings. It was done in May, June and August. The choice of research time was dictated by a seasonal increase in the number of cyclists in the road traffic. Measurements were performed on Saturday and Sunday at intersections located on major bicycle routes of a given city by means of video cameras. The total number of investigated traffic signaling cycles was 480.

The results show that app. 7% of cyclists do not respect the red signal. Moreover, it was found that as many as 95% of cyclists who disrespect the red signal enter the bicycle crossing within the first 3 s after the signal turns on (Fig. 6). It needs to be noted that all the analyzed situations occurred near crosswalks, where green signals for pedestrians and cyclists were synchronized. Cyclists knowing that their speed is higher than the speed of a pedestrian assume that they can leave the crossing in the transitory time.

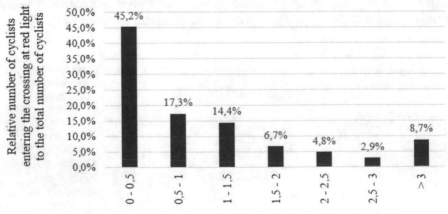

Time passed since the red light was illuminated [s]

Fig. 6. Distribution of times at an intersection after displaying the red signal for cyclists who disrespect the red signal

Cyclists were also evaluated for their age and gender. The results are presented in Figs. 7 and 8. These results indicate that male bikers 7.59% do not respect the red signal more often than women. However, female bikers who violate the red right ban account for 6.27% of bikers. The difference between women and men who enter the crossing against the red signal is smaller for the group of cyclists than for vehicle drivers. As compared to car drivers, in the group of male cyclists entered the crossing at red signal by 1.57% more often and women by 3.17% more often. The age group disrespecting the red signal ban were persons aged 19–34. In this group as many as 9.38% did not comply with the red signal. The no entry signal was most frequently respected by children under 12 ($\sim$95.5%). It should be mentioned that children under 12 were usually escorted by their parents which has a significant influence on the research results.

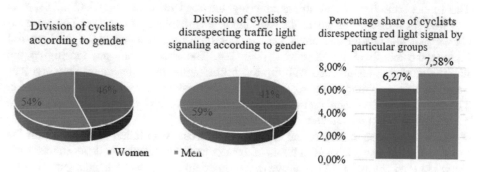

Fig. 7. Division of the bikers according to gender/division of bikers disrespecting the red signal according to gender/percentage share of non-compliance with the red signal by bikers with division into gender

Fig. 8. Division of cyclists according to age/division of cyclists who disrespect the red signal according to age according to age/percentage share of cyclists' non-compliance with the red ban with division into age

4   Non-compliance with the Red Signal by Pedestrians

Investigations of pedestrians who entered the street at the red signal were carried out at 9 pedestrian crosswalks. The crosswalks were selected so as to provide relatively high traffic volumes, different average times of waiting to cross the street and different crosswalk lengths. Age, gender and the average waiting time for red illumination were taken into consideration.

The research shows that the duration time of the red signal has a significant influence on the number of pedestrians failing to comply with it, which is closely connected with the time of waiting for the possibility to cross the street. On crosswalks where pedestrians had to wait for the red signal for a longer time the percentage of persons who violated the red signal ban was higher (Fig. 9). The longer they have to wait the more impatient they become. This in turn makes them ignore the red signal. The highest number of pedestrians entering the road on a red signal (23.67%) was reported for a crosswalk where the red signal was displayed for 120 s. The research shows that the factors that affect the number of people who enter the road on red include the length of the crosswalk (Fig. 10). The shorter they are the higher the likelihood of non-compliance with the red traffic signal by pedestrians. This involves spending a shorter time on the road and thus lower risk of being knocked by a vehicle. The shorter time they spent on the road the safer they feel so it is easier for them to take a decision to cross the street at the red signal [9, 10].

Like in the case of drivers and cyclists, pedestrians who failed to comply with the red signal were also analyzed in terms of gender and age. Results of these analyses are shown in Figs. 11 and 12. They prove that women do not respect the red signal slightly more often than men. They entered the road inappropriately in 18.78% of cases.

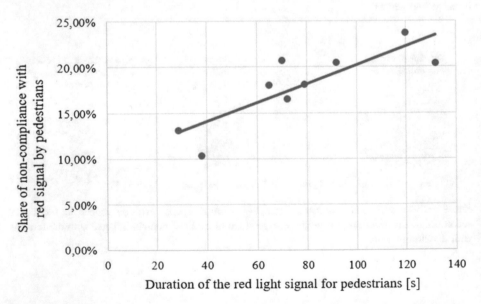

Fig. 9. Influence of the red signal duration time on non-compliance with the red by pedestrians

However, the percentage of men was 18.11%. Like in the case of cyclists, the age group of pedestrians who most often violated the traffic signaling were aged 19–34 (22.98%), whereas the most rarely - children under 12.

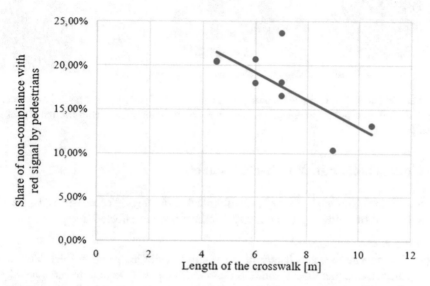

Fig. 10. Influence of the crosswalk length on the percentage share of pedestrians who fail to respect the red signal

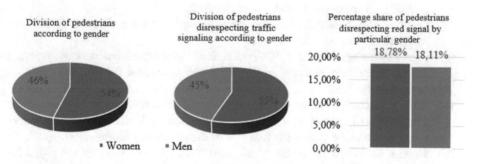

Fig. 11. Division of pedestrians according to gender/division of pedestrians failing to observe the red signal ban according to gender/percentage of non-compliance with the red signal in a given group of people

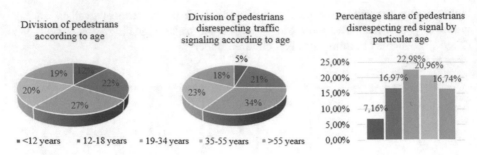

Fig. 12. Division of pedestrians according to age/division of pedestrians who fail to respect the traffic signaling according to age/percentage share of non-compliance with the red signal in a given group

5   Conclusions and Further Research

According to the research results it was found that the degree of non-compliance with the red signal of traffic signaling at city intersections are affected by:

1. In the case of car drivers:
   (a) Traffic volumes and length of car queues at the intersection inlet. The higher these traffic parameters values the more drivers who venture to run a red signal. According to the authors it is connected with an increase in the time of waiting which irritates drivers and eventually leads to an increase in road traffic risk.
   (b) Type of vehicle used by drivers. The highest percentage of motorists who disrespect the red signal was reported for motorcyclists (3.90%) and van drivers (2.77%), whereas, the lowest - buses (0.61%).
   (c) Gender. Almost two times more men disrespected the red signal (6.01%) as compared to women (3.10%).
   (d) Age. Drivers aged 35–55 (6.90%) were those who most frequently disrespected the red signal ban, whereas drivers aged <35 accounted for (3.10%).
2. In the case of cyclists:
   (a) Time passed after the red signal is illuminated. The research shows that on the average 7% of cyclists do not respect the red signal. Almost 95% of them enter a bicycle crossing within the first 3 s of the red signal illumination.
   (b) Gender. Male cyclists are the group most frequently violating the red signal (as many as ∼7.6% of them, whereas ∼6.3% of women).
   (c) Age. Cyclists at the age 19–34 entered the bike crossing at a red signal more often than cyclists of other age groups (∼9.4%). In the next group there were cyclists aged 35–55 (∼7.2%) and >55 t (∼5.3%).
3. As regards pedestrians:
   (a) The length of a red signal for pedestrians. It was found that on the average as many as 18% of pedestrians did not respect the red signal. The share of pedestrians entering a crosswalk at a red signal increases along with an increase in the red signal illumination time. According to the authors it is connected with having to wait longer for crossing the street. Like in the case of drivers, the

prospect of wasting more time leads to increased impatience and likelihood of taking more risky decisions.

(b) The length of a crosswalk. On short crosswalks pedestrians are more likely to take risk of entering the crosswalk at a red signal than on longer ones. This obviously is caused by a shorter time of being in a dangerous zone where one can be knocked by a vehicle.

(c) Age. Like in the case of cyclists, the age group of pedestrians who most frequently fail to respect the red signal are: persons aged 19–34 ($\sim$23.0%), 35–55 ($\sim$21.0%) and >55 ($\sim$16.7%).

The aim of future research will be to analyze the possibilities of reducing the number of traffic participants not applying to the red signal.

References

1. The National Road Safety Council. http://www.krbrd.gov.pl
2. Szczuraszek, T., Kempa, J., Bebyn, G., Chmielewski, J.: Road Safety Report in Bydgoszcz 2015 (2016)
3. Al-mudhaffar, A.: Impracts od Traffic Signal Control Strategies. Royal Institute of Technology, Stockholm (2006)
4. Yang, C.Y., Wassim, G.: Analysis of Red Light Violation Data Collected from Intersections Equipped with Red Light Photo Enforcement Cameras, Washington (2006)
5. Satadal, S., Subhadip, B., Nasipuri, M., Basu, D.K.: Development of an automated Red Light Violation System for Indian vehicles. In: IEEE National Conference on Computing and Communication Systems, pp. 59–64 (2009)
6. Rosenbloom, T., Wolf, Y.: Signal detection in everyday life traffic dilemmas. Accid. Anal. Prev. **34**, 763–772 (2002)
7. Moshe, B.D.: Detection of danger signals on the road and social facilitation, Bar-Ilan (2003)
8. Rosenbloom, T.: Crossing at a red light: behavior of individuals and groups. Transp. Res. Part F **12**, 389–394 (2009)
9. Hamed, M.M.: Analysis of pedestrians behavior at pedestrian crossings. Saf. Sci. **38**, 63–82 (1996)
10. Harrell, W.A.: Factors Influencing pedestrian cautiousness in crossing streets. J. Soc. Psychol. **131**, 367–372 (1991)

Assessment of the Impact Exerted by Closing Road Tunnel on Traffic Conditions on the Example of the DW902 Road in Gliwice (Poland)

Marcin Staniek[1(✉)] and Bartosz Gierak[2]

[1] Silesian University of Technology, Gliwice, Poland
marcin.staniek@polsl.pl
[2] Zarząd Dróg Miejskich, Gliwice, Poland
bartosz.gierak@zdm.gliwice.pl

Abstract. The article comments upon the impact exerted by closing the tunnel within the DW902 road on traffic conditions in the town of Gliwice. Both the tunnel in question and the town's road network have been described, and detailed information on the direct vicinity of the tunnel has been provided. The impact of the tunnel closure on traffic conditions has been analysed by comparing traffic parameters and the road situation on the day when the tunnel was closed with those of the preceding day, i.e. when the tunnel was still operational and no significant road network disturbances were observed. For purposes of the analysis of the impact exerted by the tunnel closure on traffic conditions in the town of Gliwice, individual impact zones were defined, radially extending from the tunnel location. In order to achieve the goal of the study, a comprehensive array of available measuring, computational and analytical tools were used at the Office of the Traffic Control Centre of the Municipal Road Administration in Gliwice.

Keywords: Road traffic conditions · Road traffic incidents · Road closures Road traffic control

1 Introduction

Regular operation of the transport system in urban areas is disturbed by numerous factors which one can properly aggregate [1, 2]. One group of such factors comprises various limitations of infrastructural nature [3], including efficiency of control algorithms [4, 5]. Another group is that of characteristics resulting from the modal split as well as behaviour patterns of the travelling population, and vehicle drivers in particular [6–10]. Group three is determined by disturbances stemming from the process of road pavement maintenance, i.e. repair and renovation works, covered by the road pavement management system [11]. Another group of these factors is determined by random events, e.g. accidents, traffic collisions or dynamically changing weather conditions [12].

© Springer Nature Switzerland AG 2019
G. Sierpiński (Ed.): TSTP 2018, AISC 844, pp. 116–126, 2019.
https://doi.org/10.1007/978-3-319-99477-2_11

The road traffic disturbances emerging in the transport network reduce both the capacity of its individual elements as well as the freedom of movement, which can be identified using the solutions described further on in the paper [13–16]. A consequence of the foregoing is congestion, and the resulting time losses decrease the quality of transport of passengers and goods [17–19].

2 Characterisation of the Research Area

The research area subject to the analysis addressed in the paper is the town of Gliwice, situated in the western part of the Upper Silesian Industrial Region. Compared to the rest of the region, it is a large town with the fourth highest population in the province. The town is administratively divided into 21 housing districts. Besides road transport, there is also river, air and rail transport available in the town. River transport is handled by the Gliwice Canal directly linked with the Oder. Air transport is possible via the Gliwice-Trynek airfield, assumed to ultimately function as a business type of airfield. The town's territory is cut through by several railway lines providing connection with nearby towns to the east (Zabrze, Ruda Śląska, Chorzów, Katowice) and to the west (Strzelce Opolskie, Opole).

2.1 Road Network of Gliwice

The road network of the town of Gliwice is one of the most highly developed systems in Silesian Province. The main routes cutting through the town include: the A4 motorway (linking Zgorzelec and Korczowa), the A1 motorway (linking Gdańsk and Gorzyczki), national road DK44 (linking Gliwice and Kraków), national road DK78 (linking Chałupki and Chmielnik), national road 88 (linking Strzelce Opolskie and Bytom), regional road DW408 (linking Kędzierzyn Koźle and Gliwice), regional road DW901 (linking Olesno and Gliwice) and regional road DW902, also referred to as the Urban Expressway or the DTŚ (linking Gliwice and Katowice). Besides the aforementioned roads, there are many local roads in the town, and they are very important for unobstructed transportation over the town's territory, while their traffic volume is in many cases far higher than that of regional and national roads. The most important roads of the town of Gliwice include the following streets: Kościuszki, Andersa, Chorzowska, Kujawska and Sikorskiego. The town's current road arrangement ensures efficient and quick transport within its borders. The A1 and the A4 as well as the DK88 national road perform ring road functions. It is the link between the DK88 and Rybnicka street which is assumed to ultimately perform the function of the southern ring road, and it is partially already completed within Sowińskiego street.

The main features which make it possible to influence the road traffic participants in Gliwice are traffic lights and variable message boards. Each of nearly 70 signalling units implemented in the town can operate in a cyclic mode, thus changing the parameters of the signalling program in real time based on vehicle detection. The detection is managed by means of image processing solutions (video cameras) or induction loops. A dispatcher at the Traffic Control Centre can easily and quickly adjust values of individual operating parameters of each traffic light. Moreover, every signalling unit can also operate in two basic modes, i.e. the coordinated mode, by adjusting the functioning of several programs at different intersections in order to improve traffic conditions in a selected series of streets, and the isolated mode, where a program is enabled without establishing communication with other intersections [20].

2.2 Characterisation of the Urban Expressway

The Urban Expressway (*Drogowa Trasa Średnicowa*) or the DTŚ is one of the most extensive investments in road infrastructure in the history of Silesian Province. It started as early as in 1979, while the last section was commissioned in March 2016. It cuts through several towns of the Silesian conurbation, including: Katowice, Chorzów, Ruda Śląska, Zabrze and Gliwice. It has been routed in parallel to the A4 motorway, but it features far more junctions than the latter (26 compared to 6 motorway junctions over a comparable distance). It is a three-lane road with grade-separated intersections along its entire length in each direction (with minor exceptions).

Within the borders of Gliwice, the DTŚ is divided into two sections: G1 and G2. Section G1 extends from the town border to the Kujawska street junction. It is ca. 2.8 km long and was commissioned in November 2014. Section G2 extends from the Kujawska street junction to the DK88 junction. The G2, being a section of 5.6 km, was commissioned in March 2016. There is a road tunnel along the G2, running under Dworcowa and Zwycięstwa streets, but the section also features a number of other road structures (flyovers, bridges and technological overpasses) (Fig. 1).

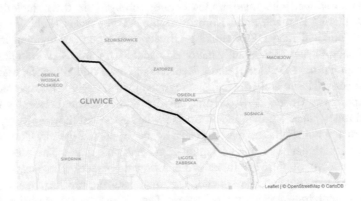

Fig. 1. Course of DTŚ road is divided into two main sections

The Gliwice road tunnel within the DTŚ is 0.493 km long, and its structure comprises two tubes: the northbound one, routed towards Wrocław, and the southbound one to Katowice. What is characteristic about the tunnel is that it features slip roads connecting to the tunnel's carriageways. By that means, it is possible to enter the north tube from Dworcowa street and exit the south tube to Dworcowa street.

2.3   Description of the Measuring Method

For purposes of the analysis of the impact exerted by the Urban Expressway tunnel closure on traffic conditions in the town of Gliwice, the area where detailed measurements were conducted has been defined. The data were acquired at intersections or roads performing significant functions from the perspective of transport within the town, and then processed and compared with data representing traffic over the town on the preceding day (an average working day without any considerable disturbances in the road network). Three descriptive measures have been proposed. The first one is traffic volume at signal-controlled intersections in a breakdown into 3 measurement zones where screening tests were performed. The measurements were conducted by means of induction loops and video detectors. The measurement area divided into individual circular zones has been depicted in Fig. 2 where the black colour marks the tunnel within the DW902 road, while red, green and blue correspond to the first, the second and the third zone, respectively.

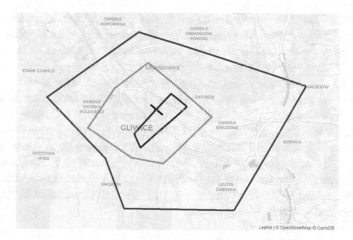

Fig. 2. Gliwice's road network divided into individual zones

Another measure of the urban road traffic description is the time needed by vehicles to travel between selected intersections. In order to acquire these data, ANPR cameras were used to capture registration numbers, as they were installed at 19 points around the town. Owing to the vehicle identification in different parts of the town, it was possible to estimate the time the driver needed to travel between specific measurement points. The last measure of traffic conditions in the town was the footage received from monitoring cameras. These devices were considered to perform an ancillary function, but they made it possible to confirm legitimacy of the data obtained in the traffic volume measurements and the number plate detection.

The tunnel was closed on 25 May 2017, from 15:10:24 to 15:28:15. The reason for the closure was a fault in the fire-protection system, forcing both tunnel tubes and both slip roads (entry and exit) to be taken out of service. As the tunnel was closed, predefined messages were automatically sent out to inform traffic participants about the recommended diversion along the entire section of the DW902 within the territory of Gliwice. Identical messages were also displayed in variable message boards at several points in the town. Instantly following the tunnel closure, dedicated procedures were triggered at several signal-controlled intersections in the closest proximity of the tunnel, making the green light signal last longer in the diversion routes. Other signal-controlled intersections were monitored, and instantaneous measures were being undertaken in response to formation of traffic jams and hold-ups.

In respect of the pre-defined research area, data concerning vehicle traffic volume and registration numbers of vehicles that used main transport routes were acquired The video footage obtained from signal-controlled intersections as well as from monitoring cameras installed along the Urban Expressway was archived.

3   Analysis of the Tunnel Closure Impact on Traffic Conditions

For the sake of the analysis, all the material was processed and compared with data acquired on the day before. The intersections and transport routes highlighted in this paper are those which revealed major deviations form typical values of traffic parameters, while those that have been skipped showed inconsiderable variations of traffic volume and travelling times, and so they have been disregarded.

Zone one covered intersections in a direct vicinity and mainly east of the tunnel. The three intersections examined in the first instance were the following: (1) intersection of Częstochowska, Jagiellońska and Mitręgi streets, (2) intersection of Jagiellońska, Dworcowa and Plac Piastów streets, and (3) intersection of Bohaterów Getta Warszawskiego and Zwycięstwa streets. They were consecutive intersections forming the diversion established for the tunnel closure period on the route towards Wrocław. The change percentage of traffic volume regarding the first intersection in the series has been summarised in Table 1.

Table 1. Analysis of intersection of Częstochowska, Jagiellońska and Mitręgi streets

Time	$\%Z_{D7}$	$\%Z_{V106}$
15:00:00	+266.7	−27.8
15:05:00	−33.3	+5.3
15:10:00	+0.0	+21.4
15:15:00	+22.2	+50.0
15:20:00	+66.7	+58.3
15:25:00	+ 39.1	+50.0
15:30:00	+175.0	+40.0
15:35:00	+133.3	+136.4
15:40:00	+250.0	+28.6
15:45:00	+375.0	+38.1
15:50:00	+40.0	+118.2
15:55:00	−22.2	+6.7
16:00:00	−50.0	+16.7

The results thus obtained imply constant increment in the number of vehicles in successive intervals as evidenced by detectors V119 and V122. These detectors counted vehicles in the road cross-section at the Urban Expressway exit and in Częstochowska street which the vehicles approached directly after departing from the exit. The traffic volume increase was noticeable for roughly an hour, which showed that it was not only experienced during the tunnel closure, but also for some time following its reopening. The foregoing was caused by vehicles which – as the tunnel was opened – already queued at the exit slip road and so were unable to return to the main road.

The change percentage of traffic volume pertaining to the intersection of Jagiellońska and Plac Piastów streets has been collated in Table 2.

Table 2. Analysis of the intersection of Jagiellońska and Plac Piastów streets

Time	$\%Z_{D7}$	$\%Z_{V106}$
15:00:00	+3.6	−11.1
15:05:00	−26.8	−4.8
15:10:00	+55.6	−9.4
15:15:00	+32.3	−10.4
15:20:00	+10.0	−2.9
15:25:00	+41.7	+41.7
15:30:00	+21.4	+65.0
15:35:00	+22.0	+25.6
15:40:00	+56.5	+16.7
15:45:00	−13.2	+23.7
15:50:00	+13.6	+13.3
15:55:00	+6.1	−2.6
16:00:00	−30.6	−4.0

The measurement results clearly imply an increase in the traffic volume reported by detector D7. This detector was responsible for counting vehicles in the cross-section of Jagiellońska street, a part of which formed a section of the diversion established for the tunnel closure time. The results provided by detector V106 which counted vehicles running in the opposite direction (Plac Piastów street) are particularly interesting. Following an initial drop, the vehicle traffic volume started to rise at 15:20. It was probably due to the fact that drivers chose an alternative detour towards Katowice. The change percentage of traffic volume pertaining to the intersection of Jagiellońska and Plac Piastów streets has been summarised in Table 3.

The measurement results show a traffic volume increase at detector V127. This unit was responsible for counting vehicles in the cross-section of Bohaterów Getta Warszawskiego street as they travelled using the diversion towards Wrocław. The number of vehicles detected at this entry branch was directly associated with the status of upstream intersections. One could notice a gradual increase in the vehicle traffic volume at consecutive intersections along the diversion route. Figure 3A and B provide sample screenshots showing comparisons of queuing lengths on the tunnel closure day A and on the preceding day B at the intersection of Jagiellońska and Plac Piastów streets.

Fig. 3. Comparison of queuing lengths on the tunnel at the measurement days

Table 3. Analysis of the intersection of Bohaterów Getta Warszawskiego and Zwycięstwa streets

Time	$\%Z_{V127}$
15:00:00	+17.2
15:05:00	−23.5
15:10:00	+36.4
15:15:00	+14.7
15:20:00	+29.6
15:25:00	+21.4
15:30:00	+32.1
15:35:00	+22.2
15:40:00	+18.9
15:45:00	+14.8
15:50:00	−21.4
15:55:00	+10.3
16:00:00	−31.4

The last intersection in zone one was that of Dworcowa, Strzody and Wyszyńskiego streets. Dworcowa is a three-lane street where the average daily traffic volume is among the highest across the whole town. Its intersection with Wyszyńskiego street is the first point which enables drivers to turn towards the very town centre by travelling south form the tunnel. Table 4 summarises the change percentage of traffic volume pertaining to this intersection.

The measurement results thus obtained imply a traffic volume increase reported by detector V111 which counted vehicles in the cross-section of the extreme right-hand lane of Dworcowa street. This lane is mainly used to turn right into Wyszyńskiego street and head downtown. The excess of vehicles observed at this point was directly caused by closing the DTŚ exit into Sienkiewicza street which performs the function of a link between the DW902 and the town centre. Around 15:45 on the tunnel closure day, one could notice a considerable traffic volume decline. Once the tunnel was reopened, vehicle traffic returned to Sienkiewicza street.

Table 4. Analysis of the intersection of Dworcowa, Strzody and Wyszyńskiego streets

Time	$\%Z_{V111}$
15:00:00	−16.3
15:05:00	+14.3
15:10:00	+25.6
15:15:00	+39.5
15:20:00	+41.7
15:25:00	+21.9
15:30:00	+12.5
15:35:00	+57.1
15:40:00	+53.6
15:45:00	−27.7
15:50:00	−17.0
15:55:00	−7.0
16:00:00	−42.6

Zone two extended over intersections mainly located in a larger distance from the tunnel, the only exception being the intersection of Jana Śliwki and Orlickiego streets comprising the DTŚ entry and exit slip roads. The measurement results imply a considerable increase in the number of vehicles at points monitored by detectors D16 and D15 which were responsible for counting vehicles in the cross-section of the exit from the DTŚ towards Katowice. The queue of vehicles recorded during the tunnel closure has been shown in Fig. 4.

Fig. 4. The queue of vehicles recorded during the tunnel closure

Another intersection directly linked with the previous one where significant changes to the vehicle traffic volume were observed was that of Orlickiego and Wyspiańskiego streets. The said considerable increase in the traffic volume was reported by detector V131 counting vehicles in the cross-section of Orlickiego street. They were leaving the DTŚ and heading for the town centre. The excess of vehicles which emerged at that point after 20 min from the tunnel closure implies that drivers chose to drive further on by the DW902 and exit into Sienkiewicza street instead of exiting earlier into Orlickiego street. The exit into Orlickiego street was chosen more frequently once the hold-up in the DTŚ grew considerably.

The traffic volume analysis pertaining to the third measurement zone revealed insignificant variations, which resulted from the distance of the relevant intersections form the tunnel and the short tunnel closure time. Having analysed the aforementioned six intersections, one could conclude that the change of traffic conditions in Gliwice during the tunnel closure depended on such factors as the distance of individual intersections from the tunnel in question and the vehicle driving direction. The excellent coordination of traffic lights that controlled Jagiellońska and Bohaterów Getta Warszawskiego streets caused that those travelling the DTŚ from Katowice towards Wrocław did not encounter any major obstacles on the diversion route. It was otherwise in the case of those travelling in the opposite direction. The low capacity of the intersection connecting Sienkiewicza street and the exit from the DW902 combined with the extensive traffic light control program coordinating Jana Śliwki and Orlickiego streets had contributed to significant deterioration of traffic conditions on the way towards Katowice.

On account of the limited capabilities of the ANPR system, reliable travel time values were only calculated for routes in the direct vicinity of the tunnel. An analysis of measurement result sheets has implied that the tunnel's impact on main north-south routes was inconsiderable. The relevant deviations form reference values came to ca. 1 min. Such a low increase in transfer times was due to the short time of the tunnel closure (20 min) and the appropriate choice of settings of traffic light control programs as well as the traffic light coordination algorithms applied in individual routes, owing to which it was possible to relieve the congestion while maintaining optimum vehicle queues at other entry roads at the same time.

4 Conclusions

Following the analysis of the measurement results, it has been established that improvement or deterioration of road traffic conditions depends on the direction of the diversion route. Vehicles headed for Wrocław did not experience major obstacles and problems caused by the tunnel closure. It should also be mentioned that enabling dedicated traffic light control programs and suitable coordination of their functioning within Jagiellońska and Bohaterów Getta streets caused that, in spite of the considerable vehicle traffic volume increase, larger hold-ups were successfully avoided. It was quite the opposite in respect of vehicles headed for Katowice. However, the problem was not in the pre-established diversion route, but in the design of the DW902 exit directly preceding the former. Consequently, the authors decided to find the critical points, i.e. locations where road traffic conditions deteriorated to a considerable degree. And so 3 such critical points were identified, namely the exit from the DW902 into Sienkiewicza street, the exit from the DW902 into Jana Śliwki and Orlickiego streets and the intersection of the DW902 and the DK88.

References

1. Galińska, B.: Multiple criteria evaluation of global transportation systems—analysis of case study. In: Sierpiński, G. (ed.) Advanced Solutions of Transport Systems for Growing Mobility, vol. 631, pp. 155–171. Springer, Cham (2017)
2. Żochowska, R.: Selected issues in modelling of traffic flows in congested urban networks. Arch. Transp. 29(1), 77–89 (2014)
3. Macioszek, E., Czerniakowski, M.: Road Traffic safety-related changes introduced on T. Kościuszki and Królowej Jadwigi Streets in Dąbrowa Górnicza between 2006 and 2015. Sci. J. Silesian Univ. Technol. Ser. Transp. 96, 95–104 (2017)
4. Małecki, K.: Graph cellular automata with relation-based neighbourhoods of cells for complex systems modelling: a case of traffic simulation. Symmetry 9(12), 322 (2017)
5. Iwan, S., Małecki, K., Stalmach, D.: Utilization of mobile applications for the improvement of traffic management systems. In: Mikulski, J. (ed.) Telematics-Support for Transport, vol. 471, pp. 48–58. Springer, Berlin (2014)
6. Sierpiński, G.: Model of incentives for changes of the modal split of traffic towards electric personal cars. In: Mikulski, J. (ed.) Telematics—Support for Transport, vol. 471, pp. 450–460. Springer, Heidelberg (2014)
7. Okraszewska, R., Nosal, K., Sierpiński, G.: The role of the polish universities in shaping a new mobility culture - assumptions, conditions, experience. In: Case Study of Gdansk University of Technology, Cracow University of Technology and Silesian University of Technology. Proceedings ICERI2014, pp. 2971–2979, Seville (2014)
8. Sierpiński, G.: Revision of the modal split of traffic model. In: Mikulski, J. (ed.) Activities of Transport Telematics, vol. 395, pp. 338–345. Springer, Heidelberg (2013)
9. Młyńczak, J., Burdzik, R., Celiński, I.: Research on vibrations in the train driver's cabin during maneuvers operations. In: Proceedings of 13th DSTA. Łódź (2015)
10. Burdzik, R., Celiński, I., Czech, P.: Optimization of transportation of unexploded ordnance, explosives and hazardous substances-vibration issues. Vibroeng. Procedia. 10, 382–386 (2016)
11. Chan, C.Y., Huang, B., Yan, X., Richards, S.: Investigating effects of asphalt pavement conditions on traffic accidents in Tennessee based on the pavement management system (PMS). J. Adv. Transp. 44(3), 150–161 (2010)
12. Golob, T.F., Recker, W.W.: Relationships among urban freeway accidents, traffic flow, weather, and lighting conditions. J. Transp. Eng. 129(4), 342–353 (2003)
13. Macioszek, E.: Analysis of significance of differences between psychotechnical parameters for drivers at the entries to one-lane and turbo roundabouts in Poland. In: Sierpiński, G. (ed.) Intelligent Transport Systems and Travel Behaviour, vol. 505, pp. 149–161. Springer, Cham (2017)
14. Małecki, K., Pietruszka, P., Iwan, S.: Comparative analysis of selected algorithms in the process of optimization of traffic lights. In: Nguyen, N., Tojo, S., Nguyen, L., Trawiński, B. (eds.) Intelligent Information and Database Systems. ACIIDS 2017. Lecture Notes in Computer Science, vol. 10192, pp. 497–506. Springer, Cham (2017)

15. Macioszek, E.: The application of HCM 2010 in the determination of capacity of traffic lanes at turbo roundabout entries. Transp. Probl. **11**(3), 77–89 (2016)
16. Turoń, K., Czech, P., Juzek, M.: The concept of Walkable City as an alternative form of urban mobility. Sci. J. Silesian Univ. Technol. Ser. Transp. **95**, 223–230 (2017)
17. Żak, J., Galińska, B.: Multiple criteria evaluation of suppliers in different industries-comparative analysis of three case studies. In: Żak, J. (ed.) Advances in Intelligent Systems and Computing, vol. 572, pp. 121–155. Springer, Cham (2017)
18. Turoń, K., Golba, D., Czech, P.: The analysis of progress CSR good practices areas in logistic companies based on reports "responsible business in poland. good practices" in 2010–2014. Sci. J. Silesian Univ. Technol. Ser. Transp. **89**, 163–171 (2015)
19. Golba, D., Turoń, K., Czech, P.: Diversity as an opportunity and challenge of modern organizations in TSL area. Sci. J. Silesian Univ. Technol. Ser. Transp. **90**, 63–69 (2016)
20. Sundar, R., Hebbar, S., Golla, V.: Implementing intelligent traffic control system for congestion control, ambulance clearance, and stolen vehicle detection. IEEE Sens. J. **15**(2), 1109–1113 (2015)

Modelling, Optimization and Evaluation in Smart Transport

Traffic Modeling in Poland at the Municipal Level. Multi-purpose Model

Piotr Rosik[(⊠)], Tomasz Komornicki, and Sławomir Goliszek

Department of Spatial Organization, Institute of Geography and Spatial
Organization Polish Academy of Sciences, Warsaw, Poland
{rosik, t.komorn, sgoliszek}@twarda.pan.pl

Abstract. The main objective of this paper is to develop passenger traffic
modeling for the entire territory of Poland at a detailed spatial scale. Analysis
embraces a large number of transport regions at the municipal level and a
network of national and regional (provincial, voivodeship) roads. For this purpose statistical data has been applied indicating local conditions related to spatial
and socio-economic structure as well as functional relations. The cognitive goal
of the study is to identify factors affecting the distribution and intensity of
passenger vehicle traffic. Such identification allows for separation of those
sections of the network, for which local socio-economic conditions and specific
functional connections determine a different modeled traffic intensity from the
actual traffic based upon the General Traffic Census measure. Modeled results
have been compared with the authentic distribution of traffic on the network of
national and regional roads in Poland in 2010. Combination of the following six
travel purposes has been taken into account: commuting to work (COM),
shopping trips (CH), commuting to university (EDU), business trips (BIZ),
visiting friends and relatives (VFR) and tourist trips (TUR). Grouping the six
travel purposes in a single multi-purpose model has proved to be successful and
led to improved fitting with the General Traffic Census results.

Keywords: Traffic modeling · Car mileage · Trip purpose split
General traffic census

1 Introduction: Purpose of the Study

Passenger road traffic modeling at the state level in Poland has not been hitherto the
subject of in-depth spatial analysis. This is mainly due to the lack of comprehensive
traffic research covering the entire territory at low aggregation level of transport
regions. Most analyzes have been conducted by the means of several thousand trials at
the level of cities or agglomerations and in recent years - also voivodeships (i.e.
administrative NTS 2 regions, see [1]). At the state level the so-called [2] developed
based on the Traffic Census in 2005 with several dozen control points, only for the
national road network. Furthermore, local socio-economic conditions as well as
functional connections were often omitted in forecasting traffic concerning regional or
state level. As a result - traffic estimations - by not considering crucial changes, such as

© Springer Nature Switzerland AG 2019
G. Sierpiński (Ed.): TSTP 2018, AISC 844, pp. 129–140, 2019.
https://doi.org/10.1007/978-3-319-99477-2_12

migration and concentration of economic potential in metropolises were significantly detached from broader social and economic context.

This paper attempts to model passenger road traffic by the means of secondary statistical data at the municipal (commune, NTS 5) level. These data present connections and socio-economic flows. Matrix data for commuters and migration (based on registration and deregistration database) were implemented in the event of two trip purposes. These data illustrate the actual relations of each unit with all others. Data concerning four remaining motivations (shopping, education, business and tourism) are based upon the production-attraction approach.

The main objective of this study is to develop passenger traffic modeling for the entire territory of Poland at a detailed spatial scale (involving a large number of transport regions at municipal level) using statistical data indicating local conditions related to spatial and socio-economic structure along with functional relations. Anticipated cognitive goal is to identify factors affecting the distribution and intensity of passenger vehicle traffic. Such identification allows for separating certain sections of the network or core-periphery pattern, for which the local socio-economic conditions and specific functional connections have been determined differently in comparison to those precisely calculated for the actual distribution of traffic on the network of state and regional level roads in Poland in 2010.

The methodological objective intends to propose a research methodology enabling forecasting traffic for the entire state on the network of national and regional roads in the form of a model, acting as an additional tool facilitating application of available secondary statistics, both in terms of production-attraction (traffic potentials) and matrix data. Authors' own model for traffic speed has been implemented including a number of factors affecting the speed of vehicles. For the purpose of calculation the VISUM software has been employed.

An exercise was performed on the potential distribution of traffic on the national and regional road networks in terms of individual six travel purposes (commuting, shopping, education, business, visiting friends/relatives and tourism), as well as aggregation of these trip motivations in the form of a multi-purpose model.

The research was carried out under the scientific grant entitled: "Comprehensive modeling of passenger road traffic in Poland with identification of its local socio-economic conditions" (KoMaR PL) conducted at the Institute of Geography and Spatial Organization Polish Academy of Sciences in 2013−2016. The project was financed by the National Science Centre funds granted on the basis of Decision No. DEC-2012/05/B/HS4/04147.

2 Study Area

The time range of the study was determined for 2010, on account of conducting the General Traffic Census (GPR) on a network of national and regional roads in that year. A considerable part of statistical material and data concerning the production-attraction approach also refers to 2010. However, matrix data for commuting pertain to 2011 and for migration - the average level for 2006 and 2009 was employed.

The spatial scope of the study concerns the road network of the entire territory of Poland (network of state and regional level roads as well as major sections of local - i.e. county and municipal - roads). The reference point for the comparative analysis of the modeled and actual traffic (GPR2010) comprised the network of national and regional roads, along which the General Traffic Census was conducted.

Majority of large traffic models that have been recently developed in Poland characterize number of transport zones oscillating around 200–300. Nevertheless, the impact of data aggregation upon the results in transport studies is of great importance [3]. In this study ten times more - exactly 2321 transport regions at municipal level have been included. In addition, the external traffic of passenger cars at 62 border points was accommodated for the means of analysis.

In the implemented traffic speed model, the speed on thirteen road categories (from local roads to motorways) was defined with the use of an original three-stage manner. Subsequently, at the fourth stage of the procedure the impact of traffic intensity on the speed change was considered. Application of an alternative approach for determining traffic speed entails indirectly considering other (apart from traffic intensity) speed limitations. The main source of data regarding the speed of passenger car traffic in 2010 was the traffic speed model developed by the IGSO PAS team for the needs of iso-chrone analysis and potential accessibility models [3–6].

This model takes into account factors that are seldom considered or even neglected in standard engineering models, however strongly affecting driving conditions and thus traffic speed. Hence, the estimated traffic speed is not "unhampered", and is intended to approximate the average possible speed, taking into account the provisions of the Highway Code, technical and functional parameters of roads and traffic conditions [5, 6]. Therefore, the passenger vehicle speed model has been developed assuming the impact of three variables on vehicle speed, i.e.: population number in a 5 km buffer around given road section, built-up area density and land relief.

3 Single-Purpose Models

The research procedure in the developed model assumes execution of a series of simulations aiming at achieving possibly most fitting model in terms of personal traffic in comparison to the results of the General Traffic Census 2010. The procedure further undertakes operation on single-purpose models for the following six selected trip motivations: commuting to work (COM), shopping trips (CH), commuting to university (EDU), business trips (BIZ), visiting friends and relatives (VFR) and tourist trips (TUR).

Single-purpose traffic modeling is a frequent procedure primarily performed in cities and agglomerations (for commuting see, for education [7] and for tourism [8]).

Sierpiński [9] has evidenced that there are clear dependences between distribution of travelling frequency (by different modes) in the function of distance and the motivation to travel. Therefore, in this paper the single-purpose models were developed by the means of different assumptions regarding production and traffic attractions, as well as various functions of distance decay (partly on the basis of a complete O-D matrix) (Table 1). The source of most data including total population number, commercial law companies, number of supermarkets, hypermarkets, department stores, population in the 19–24 age group, number of students and number of beds was the Local Data Bank of the Central Statistical Office. The procedure for selecting both production and attraction was relatively intuitive and consistent with most research on traffic conducted to date.

In order to compare data in all respects, the total "mass" volume for all transport zones was unified for both production and attraction in the case of individual trip purpose. Varied parameters for distance decay function based on a series of simulations were applied. Selection of distance decay parameter results directly from the model fitting, i.e. the parameter selection for the highest value of the coefficient of determination (R^2). Exponential distance decay function $\exp(-\beta t)$ was employed, in which selected distance decay element is time. For instance, for tourist trips, the β parameter equals 0.011552 (the most fitting model). This parameter means that the attractiveness of travel destination decreases to a half after 60 min, to 0.25 after 120 min and to 0.1 at almost 200 min.

The single-motivation models are generally well fitted to the Average Annual Daily Traffic (SDR3) in GPR2010. The coefficient of determination (R^2) ranges from 0.39 (TUR) to 0.66 (VFR). In general, apart from the TUR model, the fitting of single-motivation models ($R^2 = 0.51–0.66$) is approximate to the base model ($R^2 = 0.63$) where population is both origin and destination.

Overall, high level of fitting in the single-purpose models (apart from tourist trips) indicates concentration of various traffic types within the same road sections. This is an outcome of accumulating numerous socio-economic functions in the same urban centers, mainly voivodeship capital cities, but also partly results from strong qualitative disparities in Poland's road infrastructure. Models for shopping trips (CH) and visiting friends and relatives (VFR) also spread the traffic to regional roads of non-metropolitan areas. The commuting model underestimates the traffic in rural areas and slightly overestimates this quality in some agglomerations, which can be associated with a higher share of public transport in commuting [10]. Model for business trips reflects Poland's division into western (relatively good fitting) and eastern part (underestimation of traffic caused by relatively fewer enterprises in this area). On the other hand, certain routes to large cities such as Kraków, Poznań and Wrocław are partly overestimated.

The single-motivation models differ not only in terms of production and attraction and distance decay parameters, but also, which is in a sense the consequence of the above, the distribution of traffic in a network. The model for commuting to work is the most concentric with traffic clearly focused on the roads leading to large agglomeration and has a fractal structure. The model for commuting to universities is characterized by apparent regionalism of traffic and specific academic catchment areas, mostly restricted - with the exception of the largest academic centers - to the administrative borders of regions. The model for business trips emphasizes traffic concentration in major agglomerations and routes in between them. Modeled shopping trips as well as visiting friends and relatives also spread the traffic to other roads, including regional ones. The model for tourist trips differs significantly from the others due to the fact that the location of production and attraction is substantially distant in spatial terms.

Table 1. Model fitting in individual simulations in the base model and single-motivation simulations

Trip purpose	Traffic–generating potential		Distance decay*	R^2
	Production	Attraction		
Base model	Population (2010)	Population (2010)	15 min	0.63
Commuting (COM)	Commuting matrix (2011)		O-D matrix (without commuting exceeding 120 min)	0.65
Shopping (CH)	Population (2010)	Number of supermarkets, hypermarkets, department stores (2010)	10 min	0.60
Education (EDU)	Population 19–24 age group (2010)	Number of students (2012)	20 min	0.51
Business (BIZ)	Commercial law companies (2010)	Commercial law companies (2010)	15 min	0.62
Visiting friends and relatives (VFR)	Migration matrix (2006 and 2009 average)		O-D Matrix	0.66
Tourism (TUR)	Population (2010)	Number of beds (2010)	60 min	0.39

*Half-time values of destination attractiveness.

4   Multi-purpose Model

The multi-purpose model includes all the above described sub-models based upon the trip motivation split. The annual mileage of cars in the multi-purpose model was based on the share of travelling and average length of the trip obtained from the traffic survey developed by the Central Statistical Office (CSO) entitled [11]. Nevertheless, a number of additional assumptions have been adopted. In comparison to the examined six trip purposes of presented analysis, the Pilot study does not consider "personal needs", but instead includes the motivation of "visiting friends and relatives". Such motivation has been regarded as a personal need, while tourist trips have been classified as leisure activity.

In the modal split, it has been assumed, based on certain agglomeration models, that for the majority of trip purposes, the share of individual motorization is 70%, for commuting to school and university - 30%, and for business trips - 80%. Some values for short trips (commuting to work, commuting to school and shopping trips) are slightly higher than those resulting from the majority of traffic surveys, because the main emphasis was placed on those trips, for which the municipality border is crossed (excluding the inner city traffic, where the share of passenger cars is usually correspondingly lower) (Table 2).

Table 2. Shares of car mileage in individual motivations. Source: own elaboration based on [11]

	Trip share*	Average trip length (km)**	Transport performance	Shares based on transport work	Modal split	Shares of car mileage
Commuting	47.6	12.8	609.28	46.0%	0.70	47.5%
Education	7.8	11.7	91.26	6.9%	0.30	3.0%
Shopping	20.4	7.0	142.80	10.8%	0.70	11.1%
Business	1.0	67.2	67.20	5.1%	0.80	6.0%
Visiting friends and relatives	16.0	13.6	217.60	16.4%	0.70	17.0%
Tourism	7.2	27.4	197.28	14.9%	0.70	15.4%
All	100		1325.42	1.00	0.68	100.0%

* Based on [11], Table 3. ** Based on [11], Table 15.

In comparison to the results of the [11] this analysis assumes a much higher share of long distance trips (business trips, visiting relatives and friends), which collectively account for 38% of car mileage (whereas in the CSO analysis - business trips, personal needs and spending free time is only 24% of the total transport work). The share of commuting to work is also much lower (47.5% in relation to 60.4%, respectively). According to the authors, the share of long trips in transport performance in the CSO study is underestimated (at least in relation to inter-municipal trips).

The multi-purpose model has implemented distance decay parameters, which are identical to those best fitted to each travel purpose within the single-purpose models. The authors are aware that this procedure partly simplifies the reality and to a certain extent overestimates short trips. In further research, the distance decay parameters ought to be applied to a greater degree regarding particular trip purposes resulting from traffic analysis (ideally concerning state and regional level). The multi-purpose model proved to be relatively well fitted ($R^2 = 0.72$). The model fit is greater ($R^2 = 0.75$) in case when exclusively road sections crossing municipal borders are taken into account (Figs. 1, 2).

The coefficient of determination (R^2) is higher than in individual single-purpose models, i.e. for six trip purposes: commuting, shopping, education, business, visiting friends and relatives and tourism ($R^2 = 0.39 - 0.66$). Combination of six purposes in one model has led to improving its ultimate fit. On the other hand, the value of this coefficient (0.72) is only slightly higher while compared to certain single-motivation models, in which simulation was based on matrix data (i.e. commuting: $R^2 = 0.65$ or visiting friends and relatives: $R^2 = 0.66$). Further improvement of the model fit may require considering regional variations concerning mobility.

The model is relatively well fitted to main national roads characterized by high traffic volume. In agglomerations, overestimation of traffic above 50% occurs only in case of individual exit sections. A significant underestimation of traffic occurred in the

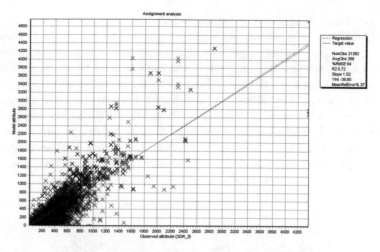

Fig. 1. Fitting of the multi-purpose model to SDR3 (GPR2010). All network sections

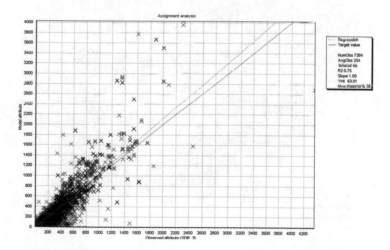

Fig. 2. Fitting of the multi-purpose model to SDR3 (GPR2010). Network sections crossing borders of municipalities

Warsaw outer bypass, which may be due to the fact that in real-life circumstances large share of drivers bypass the passage through Warsaw. Moreover, lack of sufficient number of motorways and expressways resulted in displacing the traffic to lower-level roads.

Adaptation of the model on regional roads outside agglomerations, primarily in Eastern Poland and mountain areas has lead to the underestimation of traffic in these zones.

In case of areas with relatively inconsiderable concentration of attractions (such as jobs, enterprises, tourist places, universities, etc.) in relation to their demographic potential (population density), there should be considered additional attractions at the local level for a more reliable analysis. These include local stores, churches, clinics, pharmacies, kindergartens, primary schools and other public service facilities that would act as "actual" destinations in rural areas. Factors related to the process of expanding the road network are also important. This progression was particularly meaningful in 2010. There is also a social aspect consisting in a negative perception of driving through large urban centers. Furthermore, ongoing road modernization in examined year could have resulted in the drivers' desire to seek alternative routes (Fig. 3).

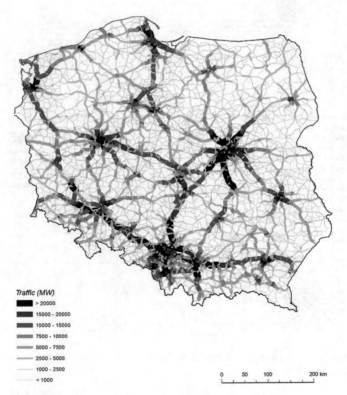

Traffic (MW)
> 20000
15000 - 20000
10000 - 15000
7500 - 10000
5000 - 7500
2500 - 5000
1000 - 2500
< 1000

0   50   100   200 km

Fig. 3. Theoretical distribution of passenger vehicle traffic in 2010. Multi-purpose model (MW)

An interesting issue is the overestimation of traffic on the routes between the Małopolskie and Śląskie as well as the Wielkopolskie and Łódzkie voivodeships (regions). In principle, traffic on all roads leading from Kraków to the Upper Silesian Industrial Region (Śląskie voivodeship) is significantly overestimated (with the exception of the undervalued regional road number 780). Another exemplification of a similar relation is provided by the A2 motorway and the parallel DK72 state road between Łódź and Konin. In this case the traffic on the motorway is also significantly overestimated. The fact that the actual traffic between closely located agglomerations (Kraków/Upper Silesian Industrial Region) or regions (Łódzkie/Wielkopolskie voivodeships) is evidently lower than that obtained in the model can be interpreted by the functioning of the so-called phantom borders, which are culturally and historically conditioned (borders of the Partitions of Poland in years 1795–1918). This can also be explained by the structural mismatch of the neighboring regional economies (e.g. between industrial Silesia and multi-functional "service-oriented" Kraków), as a result leads to the weakening of business trips volume (Fig. 4).

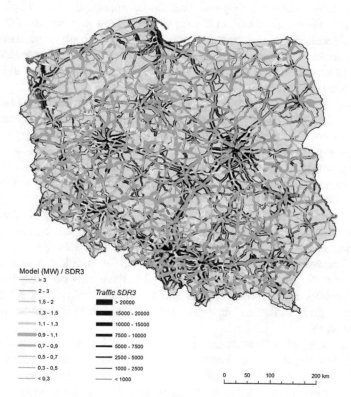

Fig. 4. Overestimation/underestimation of passenger vehicle traffic in 2010. Theoretical distribution (multi-motivation model) in relation to the GPR 2010 results.

5   Conclusions

Combination of the following six trip purposes: commuting to work (COM), shopping trips (CH), commuting to university (EDU), business trips (BIZ), visiting friends and relatives (VFR) and tourist trips (TUR) in a consolidated multi-purpose model has been successful and led to an improved fitting to the GPR 2010 results (actual traffic). The single-purpose models are also well-fitted to the Average Annual Daily Traffic (SDR3) of the GPR2010 with the coefficient of determination (R^2) ranging from 0.39 (TUR) to 0.66 (VFR). High level of fitting concerning single-motivation models (apart from tourist trips) signifies concentration of various types of traffic within the same network sections. Nevertheless, the multi-purpose model is the best fitting one ($R^2 = 0.72$, whereas considering only the sections crossing borders of municipalities - even 0.75).

The multi-purpose model is relatively well fitted in relation to actual traffic on key state-level roads characterized by high traffic volume. The exceptions are single overestimated sections of the exit roads to the largest cities (Wrocław, Poznań, Kraków and Szczecin). On the other hand, underestimation of traffic has occurred on the so-called large Warsaw bypass, which may result from significant share of individual drivers who avoid passing through the capital of Poland. Another explanation for observed traffic underestimation is the lack of a sufficient number of motorways and expressways resulted in displacing traffic into other road routes, primarily regional routes, e.g. between Warsaw and Lublin. With the expansion of infrastructure and completion of expressways and motorways, the issue of spreading traffic to lower-class roads will diminish its significance.

In order to improve the modeling process, one should seek additional attractions at local level, which act as "actual" destinations in rural areas where the concentration of modeled attractions is relatively negligible in relation to the demographic potential (population density). Inhabitants of peripheral areas are in reality mobile to a greater extent than it would result from the distribution of examined attractions. This leads to the conclusion that there are "attractions" not included in the modeling or those that are beyond statistics.

The above observations are related to the process of constructing the road network, particularly intense in year 2010, for which the study was conducted. Repair and modernization taking place in numerous road network sections may have resulted in the drivers' desire to search for alternative routes. In addition, some sections of motorways and expressways were "dead-end", i.e. there was no network effect. All mentioned network factors are temporary and after the period of network modernization one can expect "facilitated" modeling and enhanced fit of the model to the actual traffic distribution.

Implementation of comprehensive modeling for passenger car traffic after 2023, when the largest investments co-financed with the EU funds will be completed seems crucial. Then, the impact effect related to ongoing infrastructure investments on the shortest travel paths is expected to be diminishing and the planned motorway network in Poland will be almost complete. In general - until 2023 - at the state level one should anticipate: further concentration of traffic on motorways and expressways, relative stability in the modal split (and in the long-term perhaps an inversion of an unfavorable trend concerning private car prevalence), further increase of traffic in agglomerations and on main routes leading to the largest cities, where the impact radius in everyday travel will increase, mainly as an outcome of suburbanization, traffic decrease in depopulating peripheral areas and further growth in the importance of international traffic.

The study indicates that some road sections serve clearly different types (purposes) of traffic, while other sections "satisfy" the needs in terms of numerous trip purposes. This relation can be applied for instance in ranking investment projects. The demographic factor must be taken into greater consideration in transport analysis and transport policy. However, population distribution examined separately should not be the basis for investment decisions, especially in eastern Poland. The vicinity of the largest metropolises constantly requires a large investment effort in transport (not only in the field of road construction). The expected increase in traffic was underestimated when deciding on the number of lanes in case of some bypasses.

The proposed methodology may in the future become fundamental for establishing principles for forecasting traffic for the entire state's road network (by the means of assessing specific regional conditions). Knowledge of regional and local determinants is an opportunity for territorialization of transport policy. The role of developed models and the meaning of further detailed research in this field is most relevant in the area of large agglomerations and as well as peripheries, especially in the Eastern Poland. It is recommended to conduct extensive surveys, in particular concerning traffic at state level, which will take into account different aspects of traffic in rural areas.

References

1. Kulpa, T., Szarata, A.: Development of the transport model for the Masovian Voivodeship. In: Contemporary Challenges of Transport Systems and Traffic Engineering. Lecture Notes in Networks and Systems, vol. 2, pp. 193–204. Springer, Heidelberg (2017)
2. National Traffic Model elaboration by Warsaw University of Technology commissioned by GDDKiA under project 'Studium układu dróg szybkiego ruchu w Polsce' (2008)
3. Stępniak. M., Rosik P.: The impact of data aggregation on potential accessibility values. In: Geoinformatics for Intelligent Transportation. Lecture Notes in Geoinformation and Cartography, pp. 227–240. Springer, Heidelberg (2014)
4. Więckowski, M., Michniak, D., Bednarek-Szczepańska, M., Chrenka, B., Ira, V., Komornicki, T., Rosik, P., Stępniak, M., Szekely, V., Śleszyński, P., Świątek, D., Wiśniewski, R.: Road accessibility to tourist destinations of the Polish-Slovak borderland: 2010–2030 prediction and planning. Geogr. Pol. 87(1), 5–26 (2012)
5. Śleszyński, P.: A geomorphometric analysis of Poland on the basis of SRTM-3 data. Geogr. Pol. 85(4), 47–61 (2012)
6. Śleszyński, P.: Expected traffic speed in Poland using Corine land cover, SRTM-3 and detailed population places data. J. Maps 11(2), 245–254 (2015)
7. Okraszewska, R., Romanowska, A., Jamroz, K.: The effect of university campuses on the modal split of polish cities. In: Intelligent Transport Systems and Travel Behaviour. Advances in Intelligent Systems and Computing, vol. 505, pp. 65–74. Springer, Heidelberg (2017)
8. Michniak, D., Więckowski, M., Stępniak, M., Rosik, P.: The impact of selected planned motorways and expressways on the potential accessibility of the Polish-Slovak borderland with respect to tourism development. Moravian Geogr. Rep. 23(1), 13–20 (2015)
9. Sierpiński, G.: Distance and frequency of travels made with selected means of transport—a case study for the Upper Silesian conurbation (Poland). In: Intelligent Transport Systems and Travel Behaviour. Advances in Intelligent Systems and Computing, vol. 505, pp. 75–85. Springer, Heidelberg (2017)

10. Chmielewski, J., Olenkowicz-Trempała, P.: Analysis of selected types of transport behaviour of urban and rural population in the light of surveys. In: Recent Advances in Traffic Engineering for Transport Networks and System. Lecture Notes in Networks and Systems, vol. 21, pp. 27–36. Springer, Heidelberg (2018)
11. Pilot study of population communication behavior in Poland, Stage III - final report, Research work under the project "Supporting monitoring system of cohesion policy under 2007–2013 financial perspective - programming and monitoring cohesion policy in the 2014-2020 programming period", Central Statistical Office in Warsaw and the Center of Statistical Research and Education at the Central Statistical Office in Jachranka (2015) (in Polish)

Estimations of Parking Lot Capacity Using Simulations of Parking Demand as Flow of Requests for Services

Vitalii Naumov$^{(\boxtimes)}$

Transport Systems Department, Cracow University of Technology,
Krakow, Poland
vnaumov@pk.edu.pl

Abstract. Paper presents a simulation model of a parking lot operation based on the object-oriented approach. The proposed model considers stochastic nature of demand for parking services: demand is described as a couple of stochastic variables – time interval between vehicles arrival to a parking lot and a parking duration. Using Python implementation of the model, the simulation experiment was carried out in order to define functional dependence between the probability of servicing the vehicle at the parking lot and numeric parameters of demand for parking services. On the grounds of the obtained simulation results, the formula for estimations of a parking lot optimal capacity was defined.

Keywords: Parking lot capacity · Regression analysis · Python programming
Computer simulations

1 Introduction

Parking lots are the necessary elements of contemporary urban transport systems. The use of parking lots eliminates traffic jams caused by the decrease of road capacity due to vehicles parked on a roadway. Interceptive parking lots provide minimizing of traffic intensity in the central part of cities, which significantly reduce air and noise pollution. Parking lots are obligatory elements of any entertainment, office or shopping center; they enhance the attractiveness of the commercial objects for the clientele.

On the other hand, parking lots are the sources of water pollution because of their extensive impervious surfaces [1]. As a rule, parking lots need more land area than the respective office or shop buildings, this leads to the covering of large areas with asphalt and results in the excessive accumulation of heat. According to research results, shown in [2], the heat from paved areas in urban zones could even change the weather locally. It also should be mentioned, that parking lots operation demands a certain amount of funds; as objects of commercial activities, they should be profitable.

Thus, the capacity of parking lot should be sufficient to ensure the implementation of the main purpose of the parking lot as an element of transport system. At the same time, it should not be excessive in order to eliminate the negative effects of parking lots operation.

© Springer Nature Switzerland AG 2019
G. Sierpiński (Ed.): TSTP 2018, AISC 844, pp. 141–149, 2019.
https://doi.org/10.1007/978-3-319-99477-2_13

The goals of this paper are to present an object-oriented approach to simulations of a parking lot operation and to propose the method for estimation of the optimal capacity for a parking lot.

The paper has the following structure: the most relevant problems related to parking lots operation are discussed in the second section; in the third part, the mathematical model of a parking lot operation and the approach to estimating the parking lot capacity are described; the fourth section depicts a software developed by the author for simulations of a parking lot operation; the fifth part discusses the results of the simulation experiment, on which basis the dependence of the servicing probability from demand parameters and the parking lot capacity are defined; the last part offers conclusions and directions for future research.

2 Literature Review

Contemporary research projects in the field of parking systems optimization deal with the wide range of problems, such as parking lots designing [3–5], the sustainable operation of parking lots [1, 2, 6], the safety of pedestrians at parking areas [7], the use of information technologies in parking lot management [8], etc.

Parking lots design is usually discussed from the qualitative positions, not quantitative ones. R. Porter, the author of the research [8], investigates methods for maximizing the number of car parking spaces that can be placed in a car park. In his report, R. Porter assumes that parking lots design is particularly important for basement car parks in residential apartment blocks or offices, where parking spaces are being characterized by a high value. Another researcher, Ben-Joseph Eran, in his book [6] underlines an esthetic significance of parking lots design and discusses the ways to make them a natural part of sustainable living areas. In the research [7], Stark examines the physical design components of parking lots through the lenses of safety and environmental protection; he concludes his report with the statement, that while designing parking lots, it is important to consider the experience of the drivers as well as pedestrians to facilitate a safe and welcoming shared space. This conclusion has been made in the report [9] as well. The authors emphasize that every motorist must be a pedestrian before and after every trip; therefore, parking lots should be planned not just for motorists but for all users, including pedestrians.

Authors of the report [9], on the example of the Minnetonka city (Minnesota, United States) propose the reduction or even the elimination of restrictions on parking minimums (low bound restriction). It's established that parking minimums tend to be rigid, and this often provides too much parking. The research team suggests to lower minimum requirements on parking, and where applicable, to add maximums (high bound restriction). This is supposed to prevent parking underutilization while providing an alternative to meeting peak demand. It should be mentioned that the recommendations on parking restrictions, obtained by the authors of [9], are not grounded by numerical calculations, they are based mainly on literature review.

Parking demand and parking lots parameters are interdependent values. Thus, in order to determine the optimal parking strategy, it's necessary to create a model of demand forming first. The problem of travel demand modeling, where parking demand is considered as a key factor, is examined in researches [10, 11]. The authors of the paper [10] deal with enhancement of the London Transportation Studies model which is used as the major multi-modal strategic modeling tool. The proposed in [10] model contains a stand-alone component which interacts with mode and destination choice models to directly modify car demand to different zones; this model treats parking choice and capacity constraints. A mechanism, developed by authors, takes account of explicit supplies of parking spaces in all model zones in London, allowing the modeling of charge- and supply-based parking policies, as well as increasing the realism of the parking-related behavior of travelers. The paper [11] also introduces a new approach to estimation of parking demand on the grounds of numerical results for the central part of Kharkiv (Ukraine); the use of the proposed approach in transport systems planning process allows researchers to obtain more precise forecasts for demand parameters.

Existing approaches to estimation of a parking lot capacity (including the approaches, used in [10, 11]), as a rule, assume that the parking demand for the specific lot is determined and stable. Thus, the parking lot capacity in transportation planning models is usually accepted as a constant value, which does not depend on stochastic demand parameters. Authors of [12] propose to simulate a parking area in the city of Rijeka (Croatia) as a queue model with the use of MS Excel spreadsheets. They demonstrate that by applying the queuing theory methods the optimal number of servers and the required capacity in closed parking areas could be estimated.

3 Conceptual Model of a Parking Lot Operation

In this paper, a parking lot is considered as an element of city transport system; on the other hand, a lot is treated as a commercial object (as far as drivers pay a fee for the parking services). Construction of a new model aims to consider stochastic nature of demand on parking services while solving problems of parking lot parameters optimization. The problem of estimation of the parking lot optimal capacity is being resolved with the use of proposed conceptual model; however, it's not the only optimization task where the proposed model could be useful.

Demand for parking services could be characterized by the number of vehicles, parked at the lot during the given time, and by parking duration of a vehicle. For the given time period, instead of the parked vehicles number, we propose to use random time interval between arrivals of vehicles. Thus, the parking demand D_{PL} could be described by a pair of stochastic values – parking interval and parking duration:

$$D_{PL} = \left\{ \tilde{\zeta}_{PL}, \tilde{t}_{PL} \right\}, \tag{1}$$

where $\tilde{\zeta}_{PL}$ is the random variable of the time interval between arrivals of two vehicles, [sec./veh.]; $\tilde{t}_{PL}$ is the random variable of parking duration per a vehicle, [sec./veh.].

In order to measure a parking lot efficiency, we propose to use the profit of an enterprise providing the parking services. Profit as a numerical measure considers income and total operating costs, so it allows to take into account factors of different nature – economic, social, ecological parameters, etc. In general form, profit P_{PL} from a parking lot operation could be presented as follows:

$$P_{PL} = \frac{1}{3600} \cdot N_{sv} \cdot \bar{t}_p \cdot T_h - E_{PL}, \tag{2}$$

where N_{sv} is the total number of vehicles serviced at a parking lot during the given time period, [veh.]; $\bar{t}_p$ is an average parking duration per a vehicle, [sec./veh.]; T_h is a tariff per 1 h of parking per a vehicle, [€/h]; E_{PL} are total costs for a parking lot operation during the given time period, [€].

An average parking duration of a vehicle, serviced at the given parking lot, should be estimated as the expected value of a parking duration:

$$\bar{t}_p = \mu_t, \tag{3}$$

where μ_t is the expected value of the random variable $\tilde{t}_{PL}$, [sec./veh.].

The total number N_{sv} of serviced vehicles could be estimated as a product of the total number of vehicles, which arrive during the given time period T_m, [sec.], to a specific parking lot, and the probability that vehicles were serviced at this lot. Given that the total number of vehicles could be defined as a ratio between the duration of the time period T_m and the expected value of time interval between two vehicles arrived at a parking lot, a value of N_{sv} could be calculated with the use of formula (4):

$$N_{sv} = \frac{T_m}{\mu_\zeta} \cdot p_s, \tag{4}$$

where μ_ζ is the expected value of a random variable of time interval between arrivals of two vehicles in a row, [sec./veh.]; p_s is the probability that a vehicle, arrived at a parking lot, would be serviced because of enough free space in the parking lot.

Total costs of a parking lot operation could be presented in a generalized form, as a linear function from a parking lot capacity:

$$E_{PL} = c_0 + c_s \cdot C_{PL}, \tag{5}$$

where c_0 is a constant component of a parking lot operation costs, [€]; c_s are maintenance costs per one parking space, [€/unit]; C_{PL} is a parking lot capacity, [units].

The elements c_0 and c_s, in addition to operating costs, may include other components, such as social costs, environmental losses, the cost of ensuring road safety, etc.

Taking into account (3)–(5), the efficiency criterion of a parking lot operation gets the following form:

$$P_{PL} = T_m \cdot T_{1hr} \cdot \frac{\mu_t}{\mu_\zeta} \cdot p_s - c_0 - c_s \cdot C_{PL}. \tag{6}$$

Considering that the efficiency criterion (6) has its maximum value in a range of a parking lot capacity values, the optimal parking lot capacity could be determined as the root of the following equation:

$$\frac{\partial P_{PL}}{\partial C_{PL}} = 0. \tag{7}$$

It should be mentioned, that to make the expression (6) complete, a functional dependence $p_s = f\left(C_{PL}, \mu_\zeta, \mu_t\right)$ has to be determined. That could be achieved by analyzing statistical data on empirical observations results, or on the grounds of computer simulations – the approach used in this study.

4   Simulation Model of a Parking Lot

To provide a tool for identification of consistent patterns of a parking lot operation, the simulation model was developed in bounds of this research. The simulation model was implemented on the grounds of object-oriented approach with the use of Python pro-gramming language; its code could be forked from the author's GitHub page at https://github.com/naumovvs/parking-lot-model.

The proposed simulation model of a parking lot operation contains three classes, which represent base entities: *Parking* (a parking lot), *Vehicle* (a vehicle intended to park at the parking lot), and *VehiclesFlow* (demand for parking services). Additionally, the *Stochastic* class is implemented in order to simulate random parameters of the model (uniform, normal and exponential distributions are proposed within this class).

The *Vehicle* class has two fields (*apt* and *duration*) containing numeric parameters of a single request for parking services. The *apt* field contains a value of a time moment when a vehicle arrives at a parking lot, and *duration* represents a value of a time interval during which a vehicle supposed to be parked at the lot.

The *VehiclesFlow* class contains a list *vehicles* of *Vehicle* objects and *sim_duration* field representing the model period during which the flow of incoming requests should be simulated. The *vehicles* filed is a list of objects representing vehicles intended to park in the parking lot. Generation of the vehicles flow is implemented in a constructor of the *VehiclesFlow* class. The constructor arguments are the duration of model period and stochastic variables of vehicles arrival intervals and vehicles parking duration.

The *Parking* class contains a numeric field *capacity* representing the parking lot capacity, a field *demand* of the *VehiclesFlow* type, working lists *served*, *rejected*, and *parked* containing the *Vehicle* objects, the *occupancy* property, and the *simulate* method. The *demand* field describes demand for services of the parking lot. The *parked* list contains a list of the *Vehicle* objects, which describe vehicles parked at the parking lot at the current moment; this list is used for purposes of simulations. The lists *served* and *rejected* contain the *Vehicle* objects, which represent vehicle serviced at the parking lot and rejected ones respectively. The *occupancy* property returns a number of vehicles, parked at the parking lot at the current moment.

Method *simulate* of the *Parking* class directly implements the procedure of simulation of the parking lot operation process. The method has a single argument – time step for iterations of the model time. In the iterations loop, the current model time is being incremented on a value of the method's argument. Those iterations are held, if all the vehicles from the *demand* field are reviewed by the simulation procedure. If not all the vehicles were reviewed, for the first object in the list of vehicles, the following condition is being verified: if the vehicles arrival moment is less than the current model time and the parking lot occupancy is less than its capacity, then the vehicle is being added to the *parked* list; if the vehicles arrival moment is less than the current model time and the parking lot occupancy is equal to its capacity, then the vehicle is being added to the *rejected* list. If the vehicles list is empty and current occupation of the parking lot is greater than zero, the all the vehicles, which should leave the lot before the current model time, are being removed from the *parked* list.

5   Results of Experimental Studies

With the use of the developed simulation model, the full factorial experiment was conducted in order to determine the dependence of the servicing probability from demand parameters and a parking lot capacity. The probability values were estimated as a ratio of the number of served vehicles and the number of vehicles in the generated flow of requests. In the computer experiment, the requests interval is assumed to have exponential distribution and the parking duration – to be normally distributed random variable. The 24 h simulation period is considered.

The following sets of values for input parameters were used in the experiment:

- parking lot capacity (5 levels): minimum value – 10 units, maximum value – 130 units, step – 30 units;
- expected value of the vehicles arrival interval (5 levels): minimum value – 30 s, maximum value – 510 s, step – 120 s;
- expected value of the parking duration (5 levels): minimum value – 600 s, maximum value – 7800 s, step – 1800 s; the standard deviation for the parking duration is defined as 20% of the expected value in order to eliminate situations when the generated value must be truncated.

To ensure the statistical significance of the experiment, 100 runs were carried out in each series of the experiment. According to the obtained samples for each series, the sample size of 100 elements guarantees the significance level not exceeding 0.05 value.

The numerical results of the experiment show, that the dependence between probability of servicing p_s and input factor values is non-linear. Some results for the expected value of the arrivals interval in 30 s. are shown in Fig. 1.

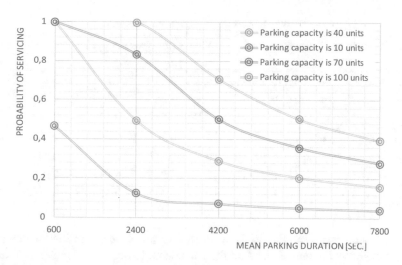

Fig. 1. Dependence of servicing probability from the expected value of parking duration (arrivals mean interval equals 30 s./veh.)

In order to define a form of the $p_s = f(C_{PL}, \mu_\zeta, \mu_t)$ functional dependence, the following hypotheses on form of the dependence were examined:

- the linear model with non-zero free coefficient: $H_1 : p_s = u_0 + a_C \cdot C_{PL} + a_\zeta \cdot \mu_\zeta + a_t \cdot \mu_t$,
- the simple linear model with the free coefficient equal to zero: $H_2 : p_s = a_C \cdot C_{PL} + a_\zeta \cdot \mu_\zeta + a_t \cdot \mu_t$,
- the power model with non-zero free coefficient: $H_3 : p_s = a_0 \cdot C_{PL}^{a_C} \cdot \mu_\zeta^{a_\zeta} \cdot \mu_t^{a_t}$,
- the power model with the free coefficient equal to one: $H_4 : p_s = C_{PL}^{a_C} \cdot \mu_\zeta^{a_\zeta} \cdot \mu_t^{a_t}$,
- the linear-logarithmic model with the free coefficient equal to zero: $H_5 : p_s = a_C \cdot \ln C_{PL} + a_\zeta \cdot \ln \mu_\zeta + a_t \cdot \ln \mu_t$,
- the mixed linear-logarithmic model with the free coefficient equal to zero: $H_6 : p_s = a_C \cdot C_{PL} + a_\zeta \cdot \ln \mu_\zeta + a_t \cdot \ln \mu_t$,
- the mixed linear-logarithmic model with the free coefficient equal to zero: $H_7 : p_s = a_C \cdot \ln C_{PL} + a_\zeta \cdot \mu_\zeta + a_t \cdot \mu_t$,
  where: a_0, a_C, a_ζ, a_t – coefficients of the regression models.

The regression coefficients for the proposed hypotheses were estimated with the use of the regression analysis tool of Microsoft Excel. The results of the completed analysis are presented in Table 1.

Table 1. Results of regression analysis.

Hypothesis on form of dependence	Values of regression coefficients				R^2
	a_0	a_C	a_ζ	a_t	value
H_1 : $p_s = a_0 + a_C \cdot C_{PL} + a_\zeta \cdot \mu_\zeta + a_t \cdot \mu_t$	0.581	$2.86 \cdot 10^{-3}$	$0.77 \cdot 10^{-3}$	$-3.25 \cdot 10^{-5}$	0.534
H_2 : $p_s = a_C \cdot C_{PL} + a_\zeta \cdot \mu_\zeta + a_t \cdot \mu_t$	0	$5.38 \cdot 10^{-3}$	$1.38 \cdot 10^{-3}$	$0.94 \cdot 10^{-5}$	0.911
H_3 : $p_s = a_0 \cdot C_{PL}^{a_C} \cdot \mu_\zeta^{a_\zeta} \cdot \mu_t^{a_t}$	0.156	0.322	0.324	-0.172	0.608
H_4 : $p_s = C_{PL}^{a_C} \cdot \mu_\zeta^{a_\zeta} \cdot \mu_t^{a_t}$	1	0.251	0.247	-0.317	0.610
H_5 : $p_s = a_C \cdot \ln C_{PL} + a_\zeta \cdot \ln \mu_\zeta + a_t \cdot \ln \mu_t$	0	0.161	0.160	-0.077	0.972
H_6 : $p_s = a_C \cdot C_{PL} + a_\zeta \cdot \ln \mu_\zeta + a_t \cdot \ln \mu_t$	0	$3.06 \cdot 10^{-3}$	0.180	-0.039	0.962
H_7 : $p_s = a_C \cdot \ln C_{PL} + a_\zeta \cdot \mu_\zeta + a_t \cdot \mu_t$	0	0.187	$0.84 \cdot 10^{-3}$	$-2.81 \cdot 10^{-5}$	0.963

As we could see, according to the estimated value of the determination coefficient, the H_5 hypothesis fits the better of all the considered hypotheses. Thus, the functional dependence of the vehicle servicing probability from a parking lot capacity and parking demand parameters should be defined as follows:

$$p_s = 0.161 \cdot \ln C_{PL} + 0.160 \cdot \ln \mu_\zeta - 0.077 \cdot \ln \mu_t. \tag{8}$$

Taking into account (8), the efficiency criterion of a parking lot operation could be presented in a following form:

$$P_{PL} = T_m \cdot T_{1hr} \cdot \frac{\mu_t}{\mu_\zeta} \cdot \left(0.161 \cdot \ln C_{PL} + 0.160 \cdot \ln \mu_\zeta - 0.077 \cdot \ln \mu_t\right) - c_0 - c_s \cdot C_{PL}. \tag{9}$$

Differentiating the expression (9), on the grounds of (7), we obtain:

$$\frac{\partial P_{PL}}{\partial C_{PL}} = 0.161 \cdot \frac{T_m \cdot T_{1hr}}{C_{PL}} \cdot \frac{\mu_t}{\mu_\zeta} - c_s = 0. \tag{10}$$

A root of the Eq. (10) in regard to C_{PL} is the optimal value of a parking lot capacity:

$$\hat{C}_{PL} = 0.161 \cdot \frac{T_m \cdot T_{1hr}}{c_s} \cdot \frac{\mu_t}{\mu_\zeta}. \tag{11}$$

where $\hat{C}_{PL}$ is the optimal capacity of a parking lot, [units].

It should be noted, that the obtained formula for estimations of the optimal parking lot capacity is statistically valid only for demand parameters bounds used in the described simulation experiment.

6 Conclusions

Contemporary approaches to estimation of a parking lot capacity, as a rule, assume that the parking demand for the specific lot is known and stable. The proposed simulation model of a parking lot operation, based on object-oriented approach, allows modeling and estimation of stochastic characteristics of the parking processes. The developed model could be expanded with additional parameters and dependencies on the grounds of proposed software. The optimal parking lot capacity formula, obtained on the basis of the conducted experiment, is recommended to be used for estimation of a parking lot capacity. This expression, however, could be clarified and supplemented with the help of the presented simulation model.

As the future direction of the research, the studies of demand for parking services on the grounds of the requests flow model should be mentioned: although the exponential distribution of the arrival intervals and normal distribution of the parking duration were used in the conducted experiment, these assumptions are not justified empirically.

References

1. Urban Stormwater Management in the United States. Research report. National Research Council, p. 513 (2008)
2. Wolf, K.L.: Trees, parking and green law: strategies for sustainability. Research report. University of Washington, p. 73 (2004)
3. Lin, X., Yuan, P.: A dynamic parking charge optimal control model under perspective of commuters' evolutionary game behavior. Phys. A **490**, 1096–1110 (2018)
4. Duan, M., Chen, G., Cao, H., Zhou, H.: Game model for balanced use of parking lots. J. Southwest Jiaotong Univ. **52**(4), 810–816 (2017)
5. Van Der Waerden, P., Janssens, D., Da Silva, A.N.R.: The influence of parking facility characteristics on car drivers' departure time decisions. Transp. Res. Procedia **25**, 4062–4071 (2017)
6. Eran, B.-J.: Re-thinking a Lot: The Design and Culture of Parking, p. 184. The MIT Press, Cambridge (2015)
7. Stark, J.A.: Parking lots: where motorists become pedestrians. Research report. State University of New York at Albany, p. 51 (2012)
8. Porter, R.: Optimisation of car park designs. Research report. University of Bristol, p. 47 (2013)
9. Christopherson, K., Shoemaker, J.M., Simon, B., Zhang, A.: Transit-Oriented Development Parking Recommendations. Resilient Communities Project Report. University of Minnesota, p. 25 (2013)
10. Hudson, M., Raha, N.: A city-wide, capacity-constrained parking choice model. In: Proceedings of the 2010 European Transport Conference (2010)
11. Horbachov, P., Naumov, V., Kolii, O.: Badania procesów parkowania w centralnej części miasta Charkowa. Zeszyty Naukowo-Techniczne SITK RP, Oddział w Krakowie, vol. 1 (100), pp. 125–134 (2013)
12. Maršanić, R., Zenzerović, Z., Mrnjavac, E.: Planning model of optimal parking area capacity. Promet Traffic Transp. **22**(6), 449–457 (2010)

Multiple Criteria Optimization for Supply Chains – Analysis of Case Study

Marcin Kiciński[1]([✉]), Piotr Witort[2],
and Agnieszka Merkisz-Guranowska[1]

[1] Division of Transport Systems, Poznan University of Technology,
Poznan, Poland
{marcin.kicinski,
agnieszka.merkisz-guranowska}@put.poznan.pl
[2] Piotr Witort Consulting, Poznan, Poland
p.witort@gmail.com

Abstract. Completing of a production process in any enterprise requires designing of a proper supply chain. Its non-compliance with the ever-changing client needs may result in a variety of problems that are perceived in different ways by different stakeholders when attempting to resolve a decision problem. The paper presents an example solution of a problem related to planning of supplies of components of a final product. The proposed non-linear, deterministic mathematical model of a decision problem includes a set of 5 criteria: costs of warehousing, transport, stock in transit and the criterion of time of transport and warehouse efficiency. Such an approach allowed including various aspects of the enterprise operation and the operation of its individual departments such as supplies department, warehouse department, marketing/sales department and management.

Keywords: Multiple criteria optimization · Mathematical modelling
Supply chains

1 Introduction

Advancing globalization of economy in combination with the advancement of information technology creates new possibilities for many production entities in terms of collaboration with other partners in the supply chain or improvement of the operation of an enterprise in the context of market competition [1–3]. The continuously observed evolution in this matter, in the beginning of the 21st century intensified in the area of management integration [4–7], risk analysis [8] or optimization of sustainable supply chains (SSC) [9]. Kumar et al. [2] indicate that in the last decade, many researchers have increasingly focused on the area of supply chain collaboration (SCC). This allows, inter alia, a reduction of the level of inventories and lead time, increases flexibility, customer satisfaction, market share or ensures a satisfactory level of profits. In the authors' opinion, due to a variety of factors having impact on the objectives of a given business enterprise, introducing changes in the supply chains requires an individual approach.

© Springer Nature Switzerland AG 2019
G. Sierpiński (Ed.): TSTP 2018, AISC 844, pp. 150–160, 2019.
https://doi.org/10.1007/978-3-319-99477-2_14

2 Optimization of Supplies in Logistic Systems

In literature of the subject, we may find a series of proposals related to the improvement of functioning of supply systems [10–17]. Its large part pertains to actions aiming at better efficiency, which is not tantamount to optimum functioning [18].

As for the proposals utilizing techniques of optimization, a good example can be the approach of Chern and Hsieh [12]. The authors proposed a multicriteria linear model to resolve the problems of main planning for a predetermined supply chain network, where the decision variable was the quantity of the finished product. In their model, they included 3 different criteria: (1) minimization of penalties for late delivery, (2) minimization of use of the outsourcing throughput and (3) minimization of costs of production, transport and warehousing. When creating the model, constraints were taken into consideration resulting from e.g. demand and capacity of the means of transport and warehouses. A similar decision variable to that proposed by Chern and Hsieh [12] was put forward by ElMaraghy and Majety [13] in the optimization of the system of supplies. The approach consisted in comparing two linear models of the same chain of supplies. The first model was optimized in terms of quality of the manufactured goods and the second additionally included the time of delivery. The performed computational experiments allowed adapting the demand for various materials on individual stages of the chain of supplies and scattering the call for materials among several suppliers. The approach of Chen et al. [11] considers only two evaluation criteria of the system of supplies: minimization of cumulative costs of production and transport and minimization of the value of the orders. In this example, the authors allowed for the non-linearity of the problem. Besides, the decision variable was also different i.e. the selection of a given supplier. A much higher decision level (strategic) is considered by the solution proposed by Bassett and Gardner [10], in which the optimized structure is the global structure of the chain of supplies for Dow AgroSciences. The authors based their model upon linear integer programming, in which the design of the chain of supplies network, the monthly production plan and the schedule of deliveries underwent parallel optimization. The criterion function was the maximization of profit. Gupta, Vanajakumari and Sriskandarajah [16] formulated the problem of optimization of supplies as a problem of developing of a schedule of distribution, in which the costs of warehousing of the distributor were minimized. The constraint adopted in the model resulted directly from the production process. According to the authors' approach, the problem was to resolve a conflict between the production plan of the plant and the schedule of dispatches from the distributor to its clients and find a compromise for both stakeholders of the chain of supplies. In another example, Grajek and Zmuda-Trzebiatowski [15] and Grajek et al. [14] proposed an optimization of the system of supplies for excise tax goods (alcoholic beverages). Owing to the tax related requirements, this type of goods requires special scrutiny in the chain of supplies. In their approach, the authors proposed a formulation of a schedule of deliveries from many suppliers to a single recipient. They analyzed the problem in the aspect of two minimized criteria i.e. scattering of the handling of the supplies (unloading) at the recipient and scattering of the handling of supplies (loading) at the suppliers. It also assumed that the deliveries are carried out with full vehicles and the suppliers delivering to the recipient are unique.

As one can observe, the approaches are mono- and multicriteria ones. The models are both linear and non-linear. One can also observe a different methodology of the search for optimum solutions, e.g. accurate methods, heuristic methods, multicriteria algorithms, genetic algorithms, ranking methods or interactive multicriteria methods.

3   Problem Description

The problem of planning of supplies in a logistic chain discussed in this paper pertains to production facilities making final goods from components supplied by different transport companies from many regions of the world. Due to the specificity of the final products, individual materials (electronic components, printed circuit boards etc.) are of different size (external dimensions, weight). The delivered materials are stored in the factory warehouses in quantities necessary to ensure the completion of individual orders. Due to the wide range of materials, it is necessary to treat the delivered materials individually in the warehouse logistics. Nevertheless, the system of warehousing and pickup of components for production is identical for each item. It is important that each ordered material from the list attributed to the final product is needed in a given quantity to manufacture that product. Therefore, the order placement procedure cannot omit any of the components.

4   Formulation of the Decision Problem

4.1   Definition of the Decision Problem

The decision problem was defined as a determination of the moment of supply of a given material being a part of the composition of an earlier defined final product ordered by the client. Individual materials are supplied by manufacturers known in advance located on different continents and delivered via air and road transport. Additionally, transport carried out by the manufacturer is also taken into account. Based on the analysis of the decision situation, the discussed problem, from the point of view of different criteria, can be: a problem of selection, a strategic problem, a deterministic problem, a problem indirectly determined in advance or a static problem.

4.2   General Information

The mathematical model of the decision problem discussed in this paper was formulated as a multicriteria combinatory optimization task. The following assumptions have been adopted:

- The finite number of components of a given final product is known
- Materials (components), from which the products are made are used for the production of only one product type
- The demand for a given product is known and constant in time

- The maximum, constant level of inventories of individual materials is known. It is scattered along the assumed time frame
- One delivery of materials must take place within four weeks
- The entire period under consideration is 52 weeks.

4.3 Decision Variables

Two types of decision variables are determined in the problem:

- Basic variable – x_{it}, defined as the moment or completion of the supply of materials i in a given week t, of the already known time frame, as expressed mathematically:

$$x_{it} = \begin{cases} > 0, & \text{if the supply of material } i \text{ is done in a week } t, \\ 0, & \text{otherwise for } i = 1, \ldots, I \text{ and } t = 1, \ldots, T. \end{cases} \quad (1)$$

- Subsidiary variable – x'_{it}, determining the frequency, with which the individual deliveries are to be carried out, as expressed mathematically:

$$x'_{it} = \begin{cases} 1, & \text{if the supply of material } i \text{ is done in a week } t, \\ 0, & \text{otherwise for } i = 1, \ldots, I \text{ and } t = 1, \ldots, T. \end{cases} \quad (2)$$

With such decision variables, their number increases drastically, depending on the adopted number of time intervals (t) and the number of materials under consideration (i).

4.4 The Family of Criterions

The proposed mathematical model of the decision problem includes a set of five criteria important from the point of view of the stakeholders from S1 to S4 of the decision problem under consideration that, in the authors' opinion include a holistic methodology generally applied in multicriteria decision aiding [19, 20].

1. **Total warehousing costs (C1)** [PLN] – Minimized (MIN). In this criterion, 4 main components were included, i.e.: cost of capital, cost of warehousing, stock handling costs (order-placing costs) and the costs of risk. Individual components of the costs were referred to a single item of a given material (component) i of the final product, thus obtaining the unit cost of warehousing (c_i^w). Given the above, the following form of the C1 criterion was adopted:

$$C1 = \sum_{i=1}^{I} \sum_{t=1}^{T} c_i^w \cdot x_{it} \cdot x'_{it} \quad (3)$$

2. **Total cost of transport (C2)** [PLN] – Minimized (MIN). In this criterion, the unit costs of transport of a given material (component) were included being a part of the final product (c_i^t). Upon obtaining of this value, offers of transport companies for individual transport routes were considered. Besides, a maximum number of units

of a given material (component) in a standard EURO pallet was considered ($\gamma_i^{\max}$). In the model, C2 is obtained from the formula:

$$C2 = \sum_{i=1}^{I} \sum_{t=1}^{T} \frac{c_i^t}{y_i^{\max}} \cdot x_{it} \cdot x_{it}'$$ (4)

3. **Total cost of stock in transit (C3)** [PLN] – Minimized (MIN). This criterion took into consideration the frozen assets in transit i.e. loss that the enterprise sustains if cash is frozen in the stock. Therefore, the following was taken into consideration: unit cost of material (component) i of a given final product (c_i), and the time of its transport (t_i) as well as the minimum annual stock return rate (provided by Warsaw Interbank Offered Rate – WIBOR) – λ. Hence, the following form of the C3 criterion was adopted:

$$C3 = \sum_{i=1}^{I} \sum_{t=1}^{T} \frac{t_i \cdot c_i \cdot \lambda}{T} \cdot x_{it} \cdot x_{it}'$$ (5)

4. **Transport time (C4)** [days] – Minimized (MIN). This criterion includes the time needed for the carriage of individual components of the final product from the supplier to the recipient (t_i). This, in turn, allows the organization of the production and warehousing processes in advance. Given the above, the criterion has the following form:

$$C4 = \frac{1}{I \cdot T} \sum_{i=1}^{I} \sum_{t=1}^{T} t_i \cdot x_{it}$$ (6)

5. **Warehouse workload (C5)** [hour] – Minimized (MIN). This criterion refers to the work efficiency of the warehouse personnel. The idea behind it is proper adaptation of the schedule of deliveries in such a way as to evenly distribute the ordered materials in time. This criterion considers: the time of the transport of pallets with all materials (components) i of the final product (t_{it}), warehouse handling (t_i^w) and minimum order quantity of the material (component) i in a given final product (MOQ_i). The following form of the C5 criterion was adopted:

$$C5 = \sum_{i=1}^{I} \sum_{t=1}^{T} \frac{x_{it} \cdot t_{it}}{MOQ_i} + \sum_{i=1}^{I} \sum_{t=1}^{T} x_{it}' \cdot \sum_{t=1}^{T} t_i^w.$$ (7)

4.5   Constraints

In the mathematical model, the following set of constraints was included:

- Within one month, at least one delivery of material (component) i of the final product must take place:

$$\bigwedge_{t=a_1=1,a_2,\,...,\,a_n\,=\,a_1(n-1)\cdot 4,\,...,\,a_{max}\,=49} \sum_t^{t+3} x_{it}' \geq 1 \tag{8}$$

- The total quantity of the supplies of individual materials $\left(\sum_{i=1}^I x_{it}\right)$ must meet the condition of demand being equal or greater than the expectations $\left(\sum_{i=1}^I P_i\right)$. Yet, the maximum quantity of material (component) i of the final product in a year $\left(\sum_{i=1}^I x_{it}\right)$ cannot exceed 125% of the anticipated demand $\left(1.25 \cdot \sum_{i=1}^I P_i\right)$, expressed as:

$$\sum_{i=1}^I P_i \leq \sum_{i=1}^I x_{it} \leq 1.25 \cdot \sum_{i=1}^I P_i \tag{9}$$

- The weekly quantity of supplies of the material (component) i of the final product $\left(\sum_t^{t+3} x_{it}\right)$ cannot exceed the maximum storage space available for this material. The storage space equals the monthly demand for a given material (component) i (x_{it}).

$$\sum_t^{t+3} x_{it} \geq x_{it} \tag{10}$$

- The quantity of supply of the material (component) i of the final product in a given month (1 month $\approx$ 4 weeks) cannot exceed its monthly consumption (w_i).

$$\bigwedge_{t=a_1=1,a_2,\,...,\,a_n=a_1(n-1)\cdot 4,\,...,\,a_{max}\,=\,49} \sum_t^{t+3} x_{it} \geq w_i \tag{11}$$

In terms of its construction, the proposed decision model is recognized as multi-criteria, nonlinear optimization problem [21]. For this reason, its resolving requires a selection of appropriate computation algorithms.

5 Case Study – EMS Factory

5.1 Entreprise Characteristics

The decision problem pertains to one of the enterprises of a larger international consortium – a company making customized electronic components for motor vehicles and electronic components for medical and industrial applications (Electronic Manufacturing Services EMS).

The business entity does not have its own transport resources and it outsources the services. The transport jobs realized from distant parts of the world (Asia, America) use air (components of small weight and size) and marine (casings and housings of final products) transport. In Europe, the company mainly relies on road transport. The entity makes sure all suppliers are located in Europe, which significantly reduces the cost of transport and frozen capital. What is more, the company tries to minimize the costs related to the warehousing (space). Limited own warehousing forces the necessity to lease additional space depending on actual needs.

The enterprise uses services of over 200 suppliers of materials and components scattered around the world and the production area is adapted to fulfill the needs of almost 20 clients. The enterprise manufactures almost 300 final products that are not based on proprietary designs but on the schematics provided by the clients. For this reason, it is impossible to individually select the component suppliers.

5.2   Stakeholders of the Decision Problem

In the enterprise under analysis, being part of a logistic chain, 4 main stakeholders of the decision problem were identified including the most vital departments of the entity: (S1) supplies, (S2) warehousing, (S3) marketing and sales and (S4) management. The individual stakeholders differ in their expectations as for the potential solutions of the problem, as presented in Table 1.

Table 1. Expectations of the stakeholders of the decision problem and their evaluation of priorities of individual criteria

Department (field)	Needs reported by the stakeholders of the decision problem	Significance of the criterion*
(S1) Supplies	Reduction of the costs of delivery Reduction of delivery time Reduction of lead time Extended terms of payment	C1(2), C2(5), C3(3), C4(4), C5(1)
(S2) Warehousing	Reduction of costs of warehousing Reduction of storage time Even distribution of materials over all warehouse locations Unification of the load units and carriers	C1(5), C2(2), C3(1), C4(3), C5(4)
(S3) Marketing and sales	Obtainment of stable and known client demand Extension of admissible lead times of the final product Obtainment of current sales forecasts Improvement of customer service quality	C1(1), C2(5), C3(4), C4(3), C5(2)
(S4) Management	Reduction of costs of frozen capital Increase of workload Maintenance of corresponding business relations Increase of production efficiency	C1(4), C2(1), C3(5), C4(2), C5(1)

*Scale from 1 to 5, where 5 denotes the most vital criterion; example: C1(2) denotes the importance of criterion 1 (total warehousing costs) = 2.

The varied demands of the stakeholders also imply different perception of individual criteria. As we can observe in Table 1, for example, criterion C1: total warehousing costs for the warehousing department is of fundamental significance – C1(5), while for the marketing and sales department it is merely – C1(1).

Nevertheless, it is noteworthy that the expectations form a certain whole aimed at a single target – the good of the client. Hence, the adopted solution must allow for a certain compromise.

5.3 Computational Experiment

The computational experiment was performed for one of the final products ordered by the client, made of 56 components, i.e.: upper and lower housing, capillaries, printed circuit boards, transistors, diodes, resistors, varistors and the main boards.

In order to calculate the necessary parameters required to carry out the computational experiment, information was collected as regards the database of suppliers of individual materials, the database of carriers, the demand for the final product, types of load units, material carriers, lead times, times of delivery from the suppliers and costs of warehousing. A fragment of the matrix of the unit costs of stock in transit for the materials (components) from 1 to 3 has been shown in Table 2.

Table 2. The matrix of the unit costs [PLN] of stock in transit for the materials (components) $i = 1, 2, 3$ of the final product

Item number (component) i	Unit costs of stock in transit in week t				
	1	2	3	4	5
1	0.00042	0.00042	0.00042	0.00042	0.00042
2	0.00057	0.00057	0.00057	0.00057	0.00057
3	0.00413	0.00413	0.00413	0.00413	0.00413

In the computational experiment, a two-stage procedure was adopted:

- Stage I: Generation of solutions using the epsilon-constraint method applied in this type of non-linear multicriteria decision problems [21–24]. To this end, Evolver tool was applied based on evolution algorithms. Eventually, 435 admissible solutions were obtained, 28 of which were pareto-optimal.
- Stage II: Evaluation and analysis of pareto-optimal solutions. At this stage, it is recommended to perform a review of the pareto-optimal solutions using, for example, 3-dimensional radial coordinate visualization (3D-RadVis) [25], Light Beam Search Method (LBS) [26] or Pareto Front Viewer (PFV) [27]. They allow a visualization of various Pareto frontiers, which is a pre-condition of efficient interactive optimization. An example of visualizations of pareto-optimal results in the LBS software for various X, Y and Z axes combinations has been shown in Fig. 1.

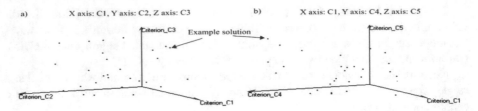

Fig. 1. Visualization of the obtained pareto-optimal results in the LBS software

Eventually, given the preferences of the stakeholders (from S1 to S2) who collectively defined the reference point, the adopted compromise solution was such having the following values of individual criteria: C1 = 22,102 PLN (1 USD = 3.45 PLN), C2 = 315,000 PLN, C3 = 10,450 PLN, C4 = 2.58 days and C5 = 12.20 h. This solution is characterized by a reduction of the frozen capital by approx. 1%, at the expense of the warehousing workload (by approx. 4 h compared to the ideal value). Besides, the warehousing costs grew slightly. This change will be practically unnoticeable for the enterprise.

6 Conclusions

As one may observe, the problem of optimization of a logistic system is complex. In the case under analysis – optimization of supplies – it requires including the opinions of many stakeholders of the decision problem. It is noteworthy that an analysis of the specificity of functioning of an enterprise resulted in the necessity of dividing the cost criterion into three components: warehousing costs, transport costs and the costs of stock in transit. The presented approach confirms the applicability of the presented methodology in resolving optimization issues in supply chains.

References

1. Daganzo, C.F.: A Theory of Supply Chains. Springer, Heidelberg (2003)
2. Kumar, G., Benerjee, R.N., Meena, P.L., Ganguly, K.K.: Joint planning and problem solving roles in supply chain collaboration. IIMB Manag. Rev. **29**, 45–57 (2017)
3. Lambert, D., Cooper, M.: Issues in supply chain management. Ind. Mark. Manag. **29**, 65–83 (2000)
4. Christopher, M.: Logistics and Supply Chain Management. Creating Value-Adding Networks. Prentice Hall, Upper Saddle River (2005)
5. Monczka, R.M., Handfield, R.B., Giunipero, L.: Purchasing and Supply Chain Management. Cengage Learning, Boston (2008)
6. Power, D.: Supply chain management integration and implementation: a literature review. Supply Chain Manag. Int. J. **10**(4), 252–263 (2005)
7. Sawik, T.: Supply Chain Disruption Management Using Stochastic Mixed Integer Programming. International Series in Operations Research & Management Science, vol. 256. Springer, New York (2018)

8. Jacyna-Gołda, I., Merkisz-Guranowska, A., Żak, J.: Some aspects of risk assessment in the logistics chain. J. KONES Powertrain Transp. **21**(4), 193–201 (2014)
9. Barbosa-Póvoa, A.P.: Optimising sustainable supply chains: a summarised view of current and future perspectives. In: Barbosa-Póvoa, A.P., Corominas, A., Miranda, J.L. (eds.) Optimization and Decision Support Systems for Supply Chains, pp. 1–11. Springer, New York (2017)
10. Bassett, M., Gardner, L.: Optimizing the design of global supply chain at Dow AgroSciences. Comput. Chem. Eng. **34**, 254–265 (2010)
11. Chen, Y.T., Che, Z.H., Chiang, T.-A., Chiang, C.J., Che, Z.-G.: Modelling and solving the collaborative supply chain planning problems. In: Chou, S., Trappey, A., Pokojski, J., Smith, S. (eds.) Global Perspective for Competitive Enterprise. Economy and Ecology Proceedings of the 16th ISPE International Conference on Concurrent Engineering, pp. 565–572. Springer, London (2007)
12. Chern, C., Hsieh, J.S.: A heuristic algorithm of master planning that satisfies multiple objectives. Comput. Oper. Res. **34**, 3491–3513 (2007)
13. ElMaraghy, H.A., Majety, R.: Integrated supply chain design using multi-criteria optimization. Int. J. Adv. Manuf. Technol. **37**(3), 371–399 (2007)
14. Grajek, M., Kiciński, M., Bieńczak, M., Zmuda-Trzebiatowski, P.: MCDM approach to the excise goods daily delivery scheduling problem. Case study: alcohol products delivery scheduling under intra-community trade regulations. Procedia Soc. Behav. Sci. **111**, 751–760 (2014)
15. Grajek, M., Zmuda-Trzebiatowski, P.: A heuristic approach to the daily delivery scheduling problem. Case study: alcohol products delivery scheduling within intra-community trade legislation. LogForum **10**(2), 163–173 (2014)
16. Gupta, S., Vanajakumari, M., Sriskandarajah, C.: Sequencing deliveries to minimize inventory holding cost with dominant upstream supply chain partner. J. Syst. Sci. Syst. Eng. **18**(2), 159–183 (2009)
17. Ruiz-Torres, A.J., Mahmoodi, F., Zeng, A.Z.: Supplier selection model with contingency planning for supplier failures. Comput. Ind. Eng. **66**, 374–382 (2013)
18. Woźniak, W., Gilewski, M.: Usprawnienie łańcucha dostaw poprzez reorganizację procesu zarzadzania zapasami na przykładzie wybranego przedsiębiorstwa. In: Patalas-Maliszewska, J., Jakubowski, J., Kłos, S. (eds.) Inżynieria produkcji: planowanie, modelowanie, symulacja, pp. 157–163. Instytut Informatyki i Zarządzania Produkcją Uniwersytetu Zielonogórskiego, Zielona Góra (2015)
19. Branke, J., Deb, K., Miettinen, K., Słowiński, R. (eds.): Multiobjective optimization: interactive and evolutionary approaches. In: State-of-the-Art Survey Series of the Lecture Notes in Computer Science, vol. 5252, Springer, Berlin (2008)
20. Halmes, Y.Y.: Harmonizing the omnipresence of MCDM in technology, society, and policy. In: Shi, Y., Wang, S., Kou, G., Wallenius, J. (eds.) New State of MCDM in the 21st Century. Selected Paper of the 20th International Conference on Multiple Criteria Decision Making 2009, pp. 13–33. Springer, Heidelberg (2011)
21. Miettinen, K.M.: Nonlinear Multiobjective Optimization. Kluwer Academic, Boston (1999)
22. Chankong, V., Haimes, Y.Y.: Multiobjective Decision Making: Theory and Methodology. North-Holland, New York (1983)
23. Ehrgott, M., Gandibleux, X.: Multiobjective combinatorial optimization—theory, methodology, and applications. In: Ehrgott, M., Gandibleux, X. (eds.) Multiple Criteria Optimization: State of the Art Annotated Bibliographic Surveys, pp. 369–444. Kluwer Academic Publishers, Boston (2002)
24. Mavrotas, G.: Effective implementation of the ε-constraint method in Multi-Objective Mathematical Programming problems. Appl. Math. Comput. **213**, 455–465 (2003)

160 M. Kiciński et al.

25. Ibrahim, A., Rahnamayan, S., Martin, M.V., Deb, K.: 3D-RadVis: Visualization of Pareto Front in Many-Objective Optimization. COIN Report Number 2016013 (2016)
26. Jaszkiewicz, A., Słowiński, R.: The "Light Beam Search" approach—an overview of methodology and applications. Eur. J. Oper. Res. **113**(2), 300–314 (1999)
27. Lotov, A.V., Bushenkov, V.A., Kamenev, G.K.: Interactive Decision Maps. Kluwer Academic Publishers, Boston (2004)

Intelligent Decision Making in Transport. Evaluation of Transportation Modes (Types of Vehicles) Based on Multiple Criteria Methodology

Barbara Galińska[✉]

Lodz University of Technology, Lodz, Poland
barbara.galinska@p.lodz.pl

Abstract. Decision making highly influences peoples' lives and their activities. Unfortunately, nowadays decision-making process is very often affected by feeling of uncertainty and risk, whereas decision problems have become increasingly complex. In these circumstances, the meaning of 'intelligence' aspect is gaining an importance as it highly enhances the possibility of making the right decision. Additionally, intelligent decision-making models are very useful in various sectors of economy, including transportation sector. The typical decision problem may be e.g. the process of evaluating and selecting transportation system, which is being defined as a set of different types of elements, relationships and processes. One of the transport's element is transport facility point - especially car fleet (different kinds of vehicles). Selection of the most desired vehicles may determine the success of the whole transportation system for the company. Therefore, the process of evaluating and selecting the used fleet should be carefully considered and based on the intelligent approach. Also, various types of tools/techniques for intelligent decision making can be used e.g. Multiple Criteria Decision Making, Group Decision Making, Artificial Neural Networks, Metaheuristic, Fuzzy Logic, Case – Based Reasoning and Expert Systems. In the case study described, the author implements MCDM Methodology (especially Electre III/IV method) in order to make the right decision during selection of the most desired variant/type of the vehicle.

Keywords: Decision making · Intelligence · Transportation
Multi-criteria decision making · Electre III/IV method

1 Introduction

One of day-to-day human activities is decision making. Some of decisions made may have little consequences and are taken relatively fast. The other have greater significance and need deeper analysis and consideration. As a premise of decision making is a selection between a few different options (alternatives, variants), such process should be analysed thoroughly with the intention of selecting the best and the most desired variant.

© Springer Nature Switzerland AG 2019
G. Sierpiński (Ed.): TSTP 2018, AISC 844, pp. 161–172, 2019.
https://doi.org/10.1007/978-3-319-99477-2_15

In order to achieve this, a detailed analysis should be undertaken. It includes designation of each variant's consequences, carrying out evaluation process and creation of ranking based on these consequences. Thus, both in theory and practice of decision making, many various tools (methods, techniques, rules) may be applied and they all support decision-making process (selection of the most desired variant). These are in particular: Multiple Criteria Methodology, Group Decision Making, Artificial Neural Networks, Metaheuristic, Fuzzy Logic, Case – Based Reasoning, Expert Systems - all enabling rational selection.

This paper is the first article of the series dedicated to the issue of intelligence and intelligent solutions in the supply chain. It was preceded by the article titled "Logistics Megatrends and Their Influence on Supply Chains" which introduced the most important logistics megatrends and their influence on supply chains changes. Therefore, this paper objective is to indicate tools/techniques which may be applied to intelligent decision making. After a detailed examination and presentation of those tools/techniques characteristics, it will be easier to prove them in practice. The empirical analysis of this paper presents selection of the most desired transportation modes (types of vehicles) for the trading company based on the multiple criteria methodology for intelligent decision making.

2   Fundamentals of Intelligent Decision Making

Decision making is being defined as a cognitive process which comprises a group of logically intertwined thinking and computational operations. They aim at solving the decision problem by selection of the most desired variant. Each decision process provides the final selection (sometimes referred to as a solution) [1].

One of the first model for decision making was proposed by Leonard Savage in 1954. The model paid particular attention to the decision maker's (DM's) feelings after making the final decision and obtaining the final results. DM is supposed to create strategies and predict potential loss in relation to optimal decision made, and also if it is possible, to predict future circumstances [2]. In fact, Savage developed the model of decision making in conditions of risk and uncertainty which also involves various consequences after making the decision and unpredictability of upcoming events [3].

Savage's theory encountered a lot of critical comments that led to development of another decision-making models. One was formulated by Simon who indicated Savage's theory limitations by developing a model of rational choice. Its concept assumes that DMs are constrained by possessed information, limited time for making a decision and their own cognitive abilities. Simon's innovative scientific work (published in 1955) introduced a new approach to decision making theory and the model was composed of 4 main stages: (1) Intelligence, (2) Design, (3) Choice, (4) Implementation (Fig. 1) [4–6].

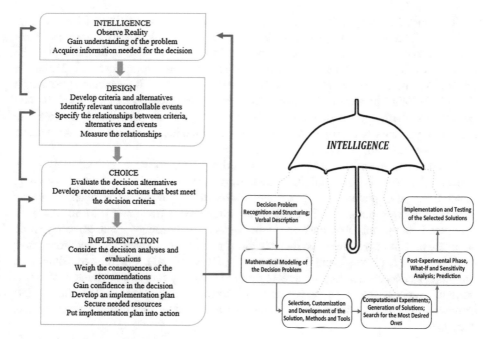

Fig. 1. Simon's model of the process of decision making (Source: [3])

Fig. 2. Żak's model of the process of decision making (Source: [7])

In Intelligence stage DM formulates decision problem and collects various kinds of information which significantly impact the decision problem. In Design stage variables (alternatives, variants) crucial for decision problem are identified, relations between them are defined and finally, evaluation criteria are indicated. Then, decision-making model for evaluation of considered alternatives is developed. In Choice stage DM evaluates alternative solutions and selects the ones which meet the evaluation criteria. In final Implementation stage, DM considers possible consequences of decisions made, develops implementation plan and collects other necessary information. Simon's model assumes that final choice is a selection from various alternatives. All model's stages must be covered by DM - there are however some open loops which allow to revert back to a previous stage or simply to start another decision-making process basing on the results from the previous implementation.

Another model, proposed by Żak, suggests different view on decision- making stages. The author also claims that 'intelligence' should be a key component in all stages of decision- making process as it mainly creates the whole process (Fig. 2) [7].

In general, above models demonstrate the increasing complexity of decision-making process. As a result, simple Decision Support Systems become insufficient and need to be replaced with the ones based on the Artificial Intelligence (AI). Additionally, AI comprises knowledge-based systems/expert systems (KBS/ES), natural language analysis, machine leaning and inference.

Thereafter, Intelligent DSS (IDSS), as the combination of AI/ES and DSS, can be seen as a DSS that provides access to data and knowledge-bases and/or conducts inference to support effective decision making in complex problem domains [1]. The research of IDSS has focused on from decision-making component expanding to integration, from quantitative model to knowledge-based decision-making approach. It all substantially increases the value of intelligent decision-making models and encourages their implementation into economy practice. Intelligent models are becoming particularly useful in many economy sectors, including transportation sector, essential for each supply chain.

In addition, one of the most significant and fastest growing branch of transport is transport by road. It is useful in almost every logistics activity. Road transport is defined as a transportation of goods and personnel by means of vehicles including private cars, light commercial vehicles and lorries with trailers [8]. Moreover, it is conducted in the unstable and uncertain logistics conditions, exposed to dynamic changes, human errors and unpredictable circumstances (e.g. weather conditions, road traffic intensity, physical and legal operational restrictions). All this contributes to 'intelligent decision-making' approach, applied to every individual element of transportation process (means of transport used, communication lanes). The 'intelligent' element in decision-making process increases the chance for making an appropriate decision.

3   Methods/Tools for Intelligent Decision Making in Transport

Present decision problems are relatively complex. It is particularly noticeable in multi-component systems such as transportation - featured with various transport facilities e.g. means of transport or traffic routes [3, 9, 10]. In order to optimize and improve the quality of decision-making processes, different methods and techniques, classified as intelligent tools are applied. The ones which are used in supply chain and its various sectors (e.g. transportation) are described below.

Multiple Criteria Decision Making (MCDM)
Described in the next section of this paper.

Group Decision Making (GDM)
GDM is the process of arriving at a judgment or a solution for a decision problem based on the input and feedback of multiple individuals. It is a group work cooperatively to achieve a satisfactory solution for the group rather than the best solution as it almost does not exist. In general, a group satisfactory solution is one that is most acceptable by the group of individuals as a whole. Since the impact of the selection of the satisfactory solution affects organisational performance, it is crucial to make the group decision-making process as efficient and effective as possible. It is therefore very important to determine what aspects make a decision making effective and how to increase the level of overall satisfaction for the solution across the group [1].

Artificial Neural Network (ANN)

ANN, usually called 'neural network' (NN), is a mathematical model or computational model that tries to simulate the structure and/or functional aspects of biological neural networks [11]. Formal neurons, the microscopic building blocks of these artificial neural networks, were introduced by McCulloch and Pitts as extremely simplified models of the biological neuron [12]. Also, they were the pioneers who initiated mathematical modelling of artificial neurons. ANNs are very suitable to solve problems that are complex, ill-defined, highly nonlinear, of many and different variables, and/or stochastic. Such problems are abundant in medicine, in finance and often in logistics [13].

Metaheuristic

The heuristic methods that can be implemented on a computer are referred to as metaheuristics [14]. They are universal computational procedures applied to solve complex optimization problems of different character and structure. Metaheuristics search over the solution space in a specific, intuitive and logical way, defined with application of common sense. Instead of reviewing all feasible solutions they screen only properly selected samples/regions of the solution space. In fact they are much faster when solving complex decision problems. At the same time they do not generate exact – optimal solutions but good approximations of such solutions. For this reason they are applied in such situations in which exact algorithms fail [7].

Fuzzy Logic (FL)

FL is a multivalued logic that defines intermediate values between conventional evaluations like true/false, yes/no, high/low, etc. [15]. The principal idea employed by FL is to allow for a partially ordered scale of truth-values, called also truth degrees, which contains the values representing false and true, but also some additional, intermediary truth degrees. That is, the set $\{0,1\}$ of truth-values of classical logic, where 0 and 1 represent false and true, respectively, is replaced in fuzzy logic by a partially ordered scale of truth degrees with the smallest degree being 0 and the largest one being 1 [16]. Such 'fuzziness' feature occurs in many real-world situations, whereby it is difficult to decide if something can be categorized exactly into a specific class or not.

Case-Based Reasoning (CBR)

CBR is an Artificial Intelligence field, based on reuse of experience. The reasoning framework used is based on the storage of experience episodes in the form of cases. Each case represents a specific situation or entity and is stored in a case library ready to be reused when new situations arrive [17]. CBR can be utilized as a decision support approach in which previous similar solutions are retrieved and consulted to solve a new problem.

Expert Systems (ES)

ES are computer-based systems that use knowledge and reasoning techniques to solve problems that would normally require human expertise. Knowledge obtained from experts and from other sources such as textbooks, journal articles, manuals and data-bases is entered into the system in a coded form. Then it is used by the system's inferencing and reasoning processes to offer advice on request [18]. ES belongs to the broader discipline of artificial intelligence (AI) which has been defined by Barr and Feigenbaum as: "the part of computer science that is concerned with designing intelligent computer systems, that is, systems that exhibit the characteristics we associate with intelligence in human behaviour - understanding language, learning, reasoning, solving problems, and so on" [19]. ES attempts to generate the best possible answer by exploring many solution paths, with using heuristic searching techniques.

4   MCDM as One of the Tool for Intelligent Decision Making

One of the tools dedicated to intelligent decision making is Multiple Criteria Decision Making (MCDM) also known as a Multicriteria Decision Aid (MCDA) or Multicriteria Decision Support, and is being defined as a field of study which derived from operational research [20, 21]. It is focused on the development and implementation of decision support tools, rules and methods in order to confront complex decision problems involving multiple criteria goals or objectives of conflicting nature [22]. It has to be though emphasized that MCDA techniques and methodologies are not just some mathematical models aggregating criteria that enable one to make optimal decisions in an automatic manner. Instead, MCDA has a strong decision support focus. In this context the DM has an active role in the decision-modelling process, which is implemented interactively and iteratively until a satisfactory recommendation is obtained and fits the preferences and policy of a particular DM or a group of DMs [3].

The methodology of MCDM/A has a universal character and can be applied in various cases when a DM solves a so called multiple criteria decision problem (MCDP) [23]. To solve this problems various methods can be used. The differences between these methodologies involve the form of the models, the model development process and their scope of application. On the basis of these characteristics, Pardalos et al. suggested the following four main streams in MCDA research [24]:

- Multiobjective mathematical programming,
- Multiattribute utility/value theory,
- Outranking relations,
- Preference disaggregation analysis.

The most popular methods implementing the outranking relations framework are the ELECTRE methods [25], as well as the PROMETHEE methods [26, 27], with different variants for addressing choice, ranking and classification problems.

In this paper the multiple criteria evaluation of transportation modes (types of vehicles) is defined as a multiple criteria ranking problem. The types of vehicles constitute the considered variants. They are evaluated by a standardized, consistent family of criteria and finally ranked from the best to the worst. The criteria evaluate various aspects of the considered variants, which are believed to be important from the perspective of the DM. To rank this vehicles the Electre III/IV method is applied.

The second group, formed by PROMETHEE methods, is one of the best known and the most popular outranking approach used in sustainable energy planning, especially in ecological transport planning [28]. Nonetheless, the research problem recognizes environmental aspect as not important one (omitted by the DM while forming the variants evaluation criteria - types of vehicles), hence the methods from this group were not applied.

Electre III/IV method is the multiple criteria method of ranking the finite set of variants which are evaluated with the application of the set of criteria [22, 29, 30]. The procedures carried out with the application of Electre III/IV method aim at the construction of preference model on the basis of pairwise comparisons of all decision variants, taking into account the thresholds which define the relation between these variants.

Computational algorithm of Electre III/IV comprises of three main stages. In the first step the definition of the set of solutions (variants) A and the consistent family of criteria F take place. Then, it is necessary to specify the value of particular criterion functions and wage indexes for each of criterion (criterion wages). Finally, the DM's preference model is defined via the thresholds of indifference qj', preference pj' and veto vj' and the importance indexes. On the second stage of the method, the consistency indexes $C(a,b)$ are computed for every pair of variants (a,b). These are presented in the form of a consistency matrix. Their values indicate to what extent the a and b are consistent with the statement that a outranks b in relation to all other criteria. Subsequently, the inconsistency index $Dj(a,b)$ is computed for each criterion j. The inconsistency index contradicts the statement that a outranks b. Finally, the outranking relation $S(a,b)$ is structured which is defined as the outranking degree $d(a,b)$ that is the aggregated measure of variants evaluation based on the consistency $C(a,b)$ and inconsistency $Dj(a,b)$ indexes. $S(a,b)$ is an overall measure which specifies to what degree a outranks b. Outranking/Reliability degrees construct the credibility matrix. On the third stage of Electre III/IV algorithm, the variants are ranked on the basis of the outranking degrees $S(a,b)$. First it is a preliminary ranking structured with the application of descending and ascending distillations which rank the variants form the best to the worst. Final, the intersection of two preorders (ascending and descending) gives the final ranking, which is usually presented in a graphical form [23].

The practical application of Electre III/IV method used as a tool for intelligent decision making in transport is described in the next section of this paper.

More information concerning MCDM methodology and its practical implementation (in regard to transportation problems) can be found in other works of the authors such as: Żak and Kiba-Janiak [31], De Brucker et al. [32], Żak [33], Żak et al. [34].

5 Application of MCDM for Intelligent Decision Making in Transportation

Major Features of the Considered Problem

The issue considered in this paper is Evaluation of Transportation Modes (Types of Vehicles) basing on the Electre III/IV method. It belongs to MCDM/A methodology and was used as a tool/technique for intelligent decision making in transportation. Evaluation of Transportation Modes was conducted for the Polish 'Multimedia' company which is one of the leading brand in electronic and household appliances. The enterprise operates on the Polish national market. It possesses 400 electronics retailers which offer among others, fridges, washing machines, cookers, kettles, televisions or laptops. Goods provided by the central warehouses of the company are delivered to the stores every day. One of the warehouse is located in Łódź (central Poland) and supports stores located within range of 250 km.

Łódź Warehouse additionally provides service for 4 electronics retailers of Multimedia, located in Poznań (western Poland). As this route belongs to the longest one (530 km distance) from logistics point of view, it was indicated by DM (Director for Logistics) for comprehensive analysis. In accordance with company's policy, deliveries to abovementioned retailers take place 5 times a week. Two kinds of vehicles operate services on this route (variants W1 and W2 described below).

On the basis of company's data analysis, within the period of July–December 2017, the company identified the problem of low delivery loads which in turn resulted in significant increase of delivery costs. In case of the analysed route an average load factor is 66%. Therefore, DM pointed the most-used vehicles for further analysis and if possible, decided to expand the scope of analysis for other options (occasional usage of variants W3 and W4). In general, the goal of the analysis is to indicate the most favourable variant which will significantly improve the load factor.

Description of the Considered Variants

Types of vehicles decision problem is formulated as a multiple criteria problem of ranking variants. Considered variants are described in Table 1.

Table 1. Variants – types of vehicles – in the decision situation.

Parameters	W1	W2	W3	W4
Total length of the vehicle [m]	8.7	8	6.7	5.5
Total width of the vehicle [m]	2.45	2.45	2.3	2.2
Length of the vehicle loading space [m]	7.3	6.4	5.2	4.2
Width of the vehicle loading space [m]	2.45	2.45	2.3	2.2
Height of the vehicle loading space [m]	2.4	2.3	2.3	2.2
Maximum mass of the vehicle [t]	12	8	5	4
Maximum cargo load compartment of the vehicle [m^3]	42.9	36.1	27.5	20.3
Maximum loading capacity [m^3]	20.0	17.0	15.0	9.4
Pallet space	18	15	12	10
The payload [t]	5	4.5	3	1.2
Fuel type	Diesel	Diesel	Diesel	Diesel
The average fuel consumption [l/100 km]	19	16	14	12
Gas emissions standards	Euro 5	Euro 5	Euro 5	Euro 5

The company is able to apply all abovementioned variants. However, analysed route is mainly operated by variant W1 (average load factor is 63%) and variant W2 (average load factor is 74%). As the load factors of space coverage remain below the acceptable level, DM decided to evaluate the variants W3 and W4 which so far were hardly ever applied. A number of criteria, important from the DM's point of view, will be taken into consideration.

Description of Evaluation Criteria

Variants evaluation criteria were formulated on the basis of the interview with the DM, his preferences and aspirations. Criteria K1–K5 are described in detail below.

K1: Cost of transport - The criterion specifies an overall unit cost of shipment transportation, expressed in [PLN/km]. The criterion was formulated on the basis of the data provided by the company and it includes the costs of vehicle's exploitation (fuel consumption) and driver's remuneration. It is the minimized criterion.

K2: Time of transport - The criterion specifies an average time necessary for delivery performance (on the analysed route). It was formulated on the basis of last 6 months period data provided by the company. It is the minimized criterion, expressed in minutes.

K3: Security of transported goods - The criterion specifies the number of pallet space. It was presumed that the way the pallets are situated in the vehicle's loading surface, has a significant impact on the security of transported goods. The greater load capacity, the lower necessity to level the transported goods (by DM indicated as an adverse factor) and as a result lower possibility of damaging the goods. The criterion is maximized.

K4: Average load factor - It is the most important variant of all evaluation criteria which specifies the average rate of maximized usage of the loading space, expressed in %. It is a maximized criterion.

K5: Payload of the vehicle - The criterion specifies the maximum vehicle's payload, expressed in tones. It was formulated on the basis of technical analysis of the data considering each variant - types of vehicles. The criterion is maximized.

Computational Experiments

In accordance with the applied Electre III/IV algorithm, the Evaluation Matrix (Table 2) was constructed, including each type of vehicle evaluation (W1–W4).

Table 2. The Evaluation Matrix of the variants – types of vehicles – in the decision situation.

Criteria	W1	W2	W3	W4
K1: Cost of transport	2.37	2.08	1.8	3.1
K2: Time of transport	570	570	510	490
K3: Security of transported goods	20	17	12	10
K4: Average load factor	63%	74%	82%	66%
K5: Payload of the vehicle	5	4.5	3	1.2

DM's preference model has been defined in the process of naming the wages of criteria and thresholds: indifference threshold q, preference threshold p and veto threshold v, which are the mode of expression DM's sensitivity to the changing value of criteria. The model has been presented in Table 3.

Table 3. The final model of preferences characteristic for the Electre III/IV method applied in the decision situation.

Preference information					
Criteria	Preference direction	Weight	Indifference threshold	Preference threshold	Veto threshold
K1	Decreasing (cost)	8	0.1	0.5	2
K2	Decreasing (cost)	5	30	60	120
K3	Increasing (gain)	7	2	5	10
K4	Increasing (gain)	9	5	10	20
K5	Increasing (gain)	3	0	1	5

In the second stage of the algorithm the outranking relation has been constructed. To build the outranking relation, the matrix of concordance and discordance were generated. On that basis, the credibility matrix has been obtained. In the final stage of the algorithm, the outranking relation S(a,b) has been applied and on the basis on the indexes of the variants and the ascending and descending distillations have been performed, formulating the structure of complete preorders. Then, they have been averaged into the median ranking and the intersection of preorders resulted in the final ranking. The results of these computations are presented in the Fig. 3.

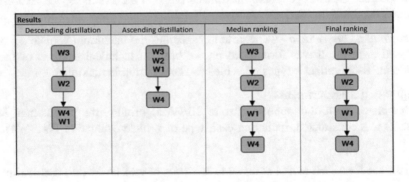

Fig. 3. The results of case study generated by a computational procedure based on the application of Electre III/IV method

Final ranking clearly indicates variant W3 as the most preferable one, outranking the other ones. The most important advantages of the vehicle W3 are: the lowest cost of transport (K1), short (but not the shortest) time of transport (K2) and finally, a high level of load factor (K4). Replacement of the currently most-used types of vehicles (variants W1 and W2) for the variant W3 will ensure 17% growth of loading space coverage (from 66% up to 83%). Also, it will directly influence costs of the transport which reduction is calculated up to 1000 Euros per month.

Application of Electre III/IV, which is one of the MCDM methods, simplified the adequate selection of the variant (type of vehicle) for the described company. The presence of 'intelligent tool' in decision process strongly influenced the decision made and its quality, enabling the DM to make the right decision.

6 Conclusions

The paper deals with the issue of intelligent decision making in transport and its practical implementation. Firstly, the description of the most important decision-making models are presented and the meaning of 'intelligent' factor is emphasized. Then, the author specifies the dynamics of the transportation process, its unstable and uncertain environment sensitive toward any sudden changes, humans' mistakes or unpredictable circumstances. These circumstances require intelligent decision making for all components of the transport.

Next part of the paper presents characteristics of the most important tools/methods for intelligent decision making in transport. Empirical part is based on one of those tools, namely MCDM. Its application supports decision-making process and selection of the most desired kind of transport (types of vehicles) and gives DM the chance for making the 'right decision'.

References

1. Lu, J., Zhang, G., Ruan, D., Wu, F.: Multi-Objective Group Decision Making: Methods, Software and Applications with Fuzzy Set Techniques. Imperial College Press, London (2014)
2. Savage, L.J.: The Foundations of Statistics. Dover Publications, New York (1954)
3. Doumpos, M., Evangelos, G.: Multicriteria Decision Aid and Artificial Intelligence: Links, Theory and Applications. Wiley, New York (2013)
4. Simon, H.A.: A behavioural model of rational choice. Quart. J. Econ. **69**(1), 99–118 (1955)
5. Simon, H.A.: The New Science of Management Decision. Prentice-Hall, Englewood Cliffs (1977)
6. Simon, H.A.: Administrative Behavior. The Free Press, New York (1997)
7. Żak, J.: The concept of intelligent decision making in logistics. In: Proceedings of CLC 2012 Conference, Jeseník, Czech Republic, 7th–9th November 2012 (2012)
8. Sierpiński, G.: Model of incentives for changes of the modal split of traffic towards electric personal cars. In: Mikulski, J. (ed.) Transport Systems Telematics 2014. Telematics - Support for Transport, vol. 471, pp. 450–460. Springer, Heidelberg (2014)
9. Okraszewska, R., Nosal, K., Sierpiński, G.: The role of the polish universities in shaping a new mobility culture - assumptions, conditions, experience. Case Study of Gdansk University of Technology, Cracow University of Technology and Silesian University of Technology. In: Proceedings of ICERI 2014 Conference, Seville, Spain, 17th–19th November 2014, pp. 2971–2979 (2014)
10. Sierpiński, G., Staniek, M., Celiński, I.: Research and shaping transport systems with multimodal travels -methodological remarks under the green travelling project. In: Proceedings of ICERI 2014 Conference, Seville, Spain, 17th–19th November 2014, pp. 3101–3107 (2014)
11. Flores, J.A.: Focus on Artificial Neural Networks. Nova Science Publishers, New York (2011)
12. McCulloch, W.S., Pitts, W.: A logical calculus of the ideas imminent in nervous activity. Bull. Math. Biophys. **5**, 115–133 (1943)
13. Graupe, D.: Principles of Artificial Neural Networks. World Scientific Publishing Co Pte Ltd., Singapore (2014)

14. Stefanoiu, D., Borne, P., Popescu, D., Filip, F., El Kamel, A.: Optimization in Engineering Sciences: Approximate and Metaheuristic Methods. Wiley, New York (2014)
15. Alavala, Ch.R.: Fuzzy Logic and Neural Networks: Basic Concepts & Application. New Age International Pvt. Ltd. (2008)
16. Belohlavek, R., Klir, G.: Concepts and Fuzzy Logic. MIT Press, Cambridge (2014)
17. Zha, X.F., Howlett, R.J. (eds.): Integrated Intelligent Systems for Engineering Design. IOS Press, Amsterdam (2006)
18. Morris, A. (ed.): The Application of Expert Systems in Libraries and Information Centres. De Gruyter, Berlin (1992)
19. Barr, A., Feigenbaum, E.A.: The Handbook of Artificial Intelligence. Morgan Kaufmann, Los Altos (1981)
20. Hillier, F., Lieberman, G.: Introduction to Operations Research. McGraw-Hill, New York (1990)
21. Żak, J.: Application of operations research techniques to the redesign of the distribution systems. In: Dangelmaier, W., Blecken, A., Delius, R., Klöpfer, S. (eds.) Advanced Manufacturing and Sustainable Logistics. Conference Proceedings of 8th International Heinz Nixdorf Symposium, IHNS 2010, Paderborn, Germany, 21th–22th April 2010 (2010)
22. Figueira, J., Greco, S., Ehrgott, M.: Multiple Criteria Decision Analysis. State of the Art Surveys. Springer, New York (2005)
23. Żak, J., Galińska, B.: Multiple criteria evaluation of suppliers in different industries-comparative analysis of three case studies. In: Żak, J., Hadas, Y., Rossi, R. (eds.) Advances in Intelligent Systems and Computing. Advanced Concepts, Methodologies and Technologies for Transportation and Logistics, vol. 572, pp. 121–155. Springer, New York (2017)
24. Pardalos, P.M., Siskos, Y., Zopounidis, C.: Advances in Multicriteria Analysis. Kluwer Academic Publishers, Dordrecht (1995)
25. Roy, B.: The outranking approach and the foundations of ELECTRE methods. Theory Decis. **31**, 49–73 (1991)
26. Brans, J.P., Mareschal, B., Vincke, P.H.: PROMETHEE: a new family of outranking methods in MCDM. In: Brans, J.P. (ed.) International Federation of Operational Research Studies (IFORS 1984), pp. 470–490. North Holland, Amsterdam (1984)
27. Brans, J.P., Vincke, P.H., Mareschal, B.: How to select and how to rank projects: the PROMETHEE method. Eur. J. Oper. Res. **24**, 228–238 (1986)
28. Wątróbski, J., Małecki, K., Kijewska, K., Iwan, S., Karczmarczyk, A., Thompson, R.G.: Multi-criteria analysis of electric vans for city logistics. Sustainability **9**(8), 1453 (2017)
29. Roy, B.: The outranking approach and the foundations of ELECTRE methods. In: Bana e Costa, C. (ed.) Readings in Multiple Criteria Decision Aid. Springer, Berlin (1990)
30. Vincke, P.: Multicriteria Decision-Aid. Wiley, New York (1992)
31. Żak, J., Kiba-Janiak, M.: A methodology of redesigning and evaluating medium-sized public transportation systems. In: Żak, J., Hadas, Y., Rossi, R. (eds.) Advances in Intelligent Systems and Computing. Advanced Concepts, Methodologies and Technologies for Transportation and Logistics, vol. 572, pp. 73–102. Springer, New York (2017)
32. De Brucker, K., Macharis, C., Verbeke, A.: Multi-criteria analysis in transport project evaluation: an institutional approach. Eur. Transp./Trasporti Europei **47**, 3–24 (2011)
33. Żak, J.: The methodology of multiple criteria decision making/aiding as a system-oriented analysis for transportation and logistics. In: Świątek, J., Tomczak, J. (eds.) Advances in Systems Science, vol. 539, pp. 265–284. Springer, New York (2017)
34. Żak, J., Redmer, A., Sawicki, P.: Multiple objective optimization of the fleet sizing problem for road freight transportation. J. Adv. Transp. **45**(4), 321–347 (2011)

Results of Research of the Traffic Safety at Signalized Intersection with Countdown Timers

Aleksander Sobota[✉], Renata Żochowska, Grzegorz Karoń,
and Marcin Jacek Klos

Faculty of Transport, Silesian University of Technology, Katowice, Poland
{aleksander.sobota, renata.zochowska, grzegorz.karon,
marcin.j.klos}@polsl.pl

Abstract. The article is a continuation of the research work carried out by the authors on determining the impact of the countdown timers on traffic conditions, and—in this case—on traffic safety at the intersections. The paper presents the results of research for assessing the impact of the role of the intersection with the countdown timers in transportation network of the city on the safety of the intersections. Moreover, the article also demonstrates the force of the influence of the countdown timers on the level of traffic safety, depending on the function of the intersection in transportation network. For scientific purposes, the traffic measurements were conducted which have been described in detail in another article [1].

Keywords: Countdown timers · Traffic safety · Function of intersection
Signalized intersection · Multilane intersection

1 Introduction

The use of the countdown timers in traffic lights control raises much controversy in Poland. In papers [2, 3], the opinions of vehicle drivers and unprotected road users have been presented, which show that these devices assist in the safe passing through the intersection. However, some representatives of the environment of the traffic engineers in Poland have a quite different view. In addition, the use of these devices in Poland is not consistent with Regulation [4].

Both in the world and in Poland the research to find solutions that help to improve road safety is being conducted. Research topics include—but are not limited to—the analyzes of road accidents [5–7], the improvement of methodology for assessing of road infrastructure elements [8] or the identification as well as estimation of the impact of the most important determinants of road safety [9]. The paper deals with one of the major safety phenomenon, frequently occurring at the signalized intersections and known as red light running (RLR). RLR-related crashes are much more likely to cause an injury or a fatality than other junction crashes. It is widely describe in worldwide literature [10, 11].

© Springer Nature Switzerland AG 2019
G. Sierpiński (Ed.): TSTP 2018, AISC 844, pp. 173–183, 2019.
https://doi.org/10.1007/978-3-319-99477-2_16

174   A. Sobota et al.

The results of the studies on the influence of the countdown timers on traffic conditions and on traffic safety in Poland has been presented in [2]. Figure 1 illustrates the variability of the number of road collision and traffic accidents in the years 2008–2012 at the selected intersections in Toruń. In 2010 the countdown timers have been installed, so this comparison shows the number of collision and accidents in the period beginning two years before the installation and ending two years after. Ranges of values corresponding to the number of collisions and the number of accidents for individual objects are different, therefore on the charts in Fig. 1 diverse scales have been used.

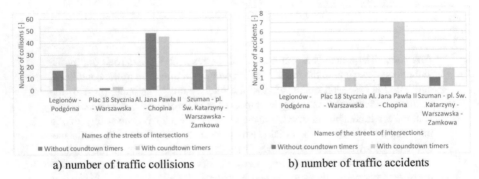

a) number of traffic collisions   b) number of traffic accidents

Fig. 1. Number of road collisions and number of traffic accidents in 2008–2012 at the selected intersections in Toruń (source: own on the basis of [2])

In the world literature other research on the functioning of the countdown timers has been presented. For example, in paper [12] the relationship between the application of the countdown timers and distribution of the length of the queue of vehicles has been determined. However, the results of any studies assessing the impact of the use of the countdown timers on the level of traffic safety, according to the function of crossing roads, have not been met. Therefore, this subject has been taken by the authors of the article as the research problem.

2   The Moment of Passing Through the Stop Line While Displaying the Red Signal – Test Results

In order to investigate the influence of the countdown timers on the level of traffic safety of intersections of different types of traffic, the numbers of vehicles and the moments of passing through the stop line when the red signal was displayed—both at the beginning and at the end of the signal—have been recorded.

2.1 Results of the Measurements for the Intersection with Urban Traffic - Intersection of De'Gaullea and Roosevelt Streets in Zabrze

Figure 2 shows the share of the vehicle passing through the stop line when the red signal has been displayed relative to the observed traffic intensity at the intersection in Zabrze.

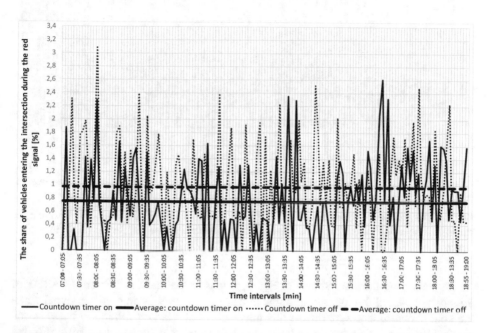

Fig. 2. Comparison of the shares of vehicles entering the intersection during the red signal for the measurement period with the countdown timers switched on and off for the object located in Zabrze

During the study, it was observed that approximately 0.8% of drivers had passed through the stop line during the red signal when the countdown timers were on and about 1% - when they were off. The difference in average number of the passes through during the red signal was 0.22% per single measurement interval (5 min). As you can see, vehicle drivers were more likely to enter the intersection at the red light while the countdown timers were on.

2.2 Results of the Measurements for the Intersection with Mixed Traffic - Intersection of Obrońców Stalingradu, Mieszka I and Jagiellonów Streets in Opole

Figure 3 shows the share of the number of the vehicle passing through the stop line in the red signal in relation to the observed traffic intensity at the intersection in Opole.

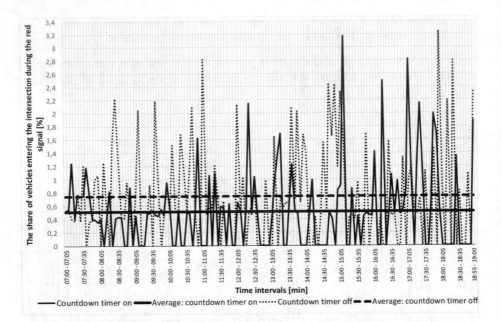

Fig. 3. Comparison of the shares of vehicles entering the intersection during the red signal for the measurement period with the countdown timers switched on and off for the object located in Opole

By analyzing the chart shown in Fig. 3, it was noted that a maximum of 3.25% of drivers had passed through the stop line in a single interval during the red signal when the countdown timers were off. At this intersection, the largest difference between the maximum shares of passes through the stop line in the red signal was 2.46% of traffic intensity in a single measurement interval.

In addition, one should notice a large difference—depending on the time of day—in the shares of vehicles entering the intersection during the red signal between the cases when the countdown timers were on and the cases when the devices were off. In the morning, i.e. from 8:00 am to 11:00 pm, the disabled countdown timers influenced on increasing the average number of passes through the stop line during the red light by 0.43% per measurement interval. By contrast, in the afternoon, i.e. from 15:00 to 18:00, the switching off the countdown timer resulted in a decrease in the number of the vehicle passing through at the red light by 0.16%. This might be due to the specifics of behaviors of the vehicle drivers, which were more in a hurry in the morning than in the afternoon, or to the lighting in the early morning, which has hindered the correct interpretation of the signal displayed.

Overall, during the study, it was observed that approximately 0.6% of drivers had entered the intersection during the red signal when the countdown timers were on and about 0.8% of the drivers - when they were off.

2.3 Results of the Measurements for the Intersection with Non Urban Traffic - Intersection of Karkonoska, Zwycięska and Jeździecka Streets in Wrocław

Figure 4 shows the share of the number of passes through the stop line during the red light compared to the observed traffic intensity at the intersection in Wrocław.

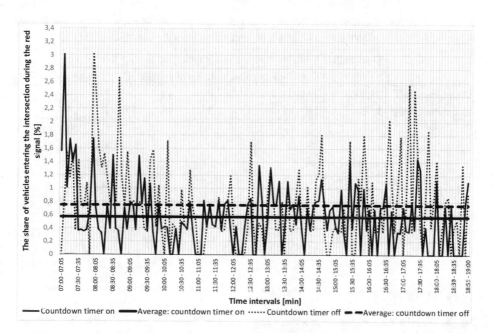

Fig. 4. Comparison of the shares of vehicles entering the intersection during the red signal for the measurement period with the countdown timers switched on and off for the object located in Wrocław

In Wrocław, it was observed that about 0.55% of drivers had passed through the stop line during the red signal when the countdown timers were on and about 0.75% of the drivers – when the devices were off. As you can see, the vehicle drivers were more likely to enter the intersection at the red signal while the timers were operating. In addition, one should notice a similar maximum value for the share of vehicles passing through the stop line at the red light regardless of the performance of the countdown timers.

3 Comparison of Results

In order to compare the impact of the use of the countdown timers on the level of traffic safety, depending on the function of the intersection, in Tables 1, 2 and 3, data on the number of entering the intersection by particular types of vehicle during the red signal with the countdown timers switched on and off has been presented.

At the intersection in **Zabrze**, the stop line has been passed during the red signal by 65 vehicles more in case, when the countdown timers were switched on. The biggest percentage difference was observed for buses. They entered the intersection during the red signal five times more in the case of disabled devices. Discussed difference has been presented in Fig. 5.

Table 1. Number of vehicles [veh/60 h] passing through the stop line when displaying the red signal with the countdown timers switched on and off at the intersection in Zabrze

	Total traffic volumes	Vehicles entering the intersection during the red signal					
		Car	Delivery truck	Minibus	Truck	Bus	Motor and bike
Countdown ON	32,560	233	14	1	3	3	1
Countdown OFF	32,361	288	11	3	3	15	0

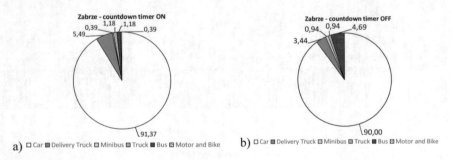

a) □ Car ■ Delivery Truck □ Minibus ▨ Truck ■ Bus ▨ Motor and Bike   b) □ Car ■ Delivery Truck □ Minibus ▨ Truck ■ Bus ▨ Motor and Bike

Fig. 5. The percentage share of particular groups of vehicles passing through the stop line while displaying the red signal with the countdown timers switched on and off at the intersection in Zabrze

In both study periods, the stop line during the red signal has been passed mostly by the passenger cars. The share of this group of vehicles exceeded 90%. With the countdown timers switched on, the entries into the intersection in the red light by delivery trucks (5.49%) have been observed, while with the countdown timers switched off - by buses (4.69%) and delivery trucks (3.44%).

Table 2 shows the number of vehicles in each of the groups passing through the stop line while displaying the red signal at the intersection in **Opole**. In the measurement period with disabled countdown timers 51 illegal entries into the intersection

Table 2. Number of vehicles [veh/60 h] passing through the stop line when displaying the red signal with the countdown timers switched on and off at the intersection in Opole

	Total traffic volumes	Vehicles entering the intersection during the red signal					
		Car	Delivery truck	Minibus	Truck	Bus	Motor and bike
Countdown ON	26,741	116	11	15	4	1	0
Countdown OFF	26,316	164	9	10	13	2	0

more than in the study period with the devices switched on have been observed. The biggest percentage difference has been noted for minibuses and heavy traffic. In addition, during the measurement period with the countdown timers off, about three times more trucks have passed through the stop line during the red signal.

Figure 6 shows a comparison of the percentage share of the passes through the stop line during the red signal by certain types of vehicles at the intersection in Opole. The share of passenger cars in the two study periods represents more than 90% of vehicles passing through the stop line at the red signal.

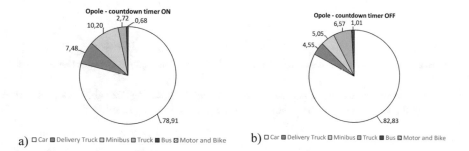

a) ☐ Car ☐ Delivery Truck ☐ Minibus ☐ Truck ■ Bus ☒ Motor and Bike   b) ☐ Car ☐ Delivery Truck ☐ Minibus ☐ Truck ■ Bus ☒ Motor and Bike

Fig. 6. The percentage share of particular groups of vehicles passing through the stop line while displaying the red signal with the countdown timers on and off at the intersection in Opole

Table 3 shows the number of vehicles passing through the stop line at the red signal at the intersection in **Wrocław**. The total number of vehicles passing through the stop line during the red signal is of greater importance (up to 366 vehicles) in the case when the countdown timers were switched off. This is particularly evident for the passenger cars and delivery trucks. Moreover, in each group of vehicles a higher number of passes through during the red light in the case when the timers disabled has been reported.

Table 3. Number of vehicles [veh/72 h] passing through the stop line when displaying the red signal with the countdown timers switched on and off at the intersection in Wrocław

	Total traffic volumes	Vehicles entering the intersection during the red signal					
		Car	Delivery truck	Minibus	Truck	Bus	Motor and bike
Countdown ON	53,354	278	17	6	2	3	0
Countdown OFF	53,769	617	35	9	3	7	1

Figure 7 shows a comparison of the percentage share of the passes through the stop line during the red signal by certain types of vehicles at the intersection in Wrocław. In the two measurement periods, the largest number of illegal entries into the intersection—except the passenger cars—has been recorded for trucks. The share of passes through the stop line at the red signal by the remaining groups of vehicles was not greater than 2%.

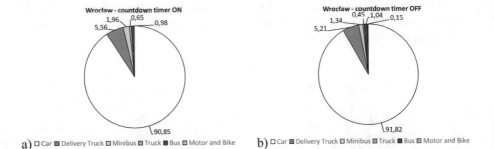

a) □ Car ■ Delivery Truck □ Minibus ▧ Truck ■ Bus ▨ Motor and Bike   b) □ Car ■ Delivery Truck □ Minibus ▧ Truck ■ Bus ▨ Motor and Bike

Fig. 7. The percentage share of particular groups of vehicles passing through the stop line while displaying the red signal with the countdown timers switched on and off at the intersection in Wrocław

Table 4 and Fig. 8 show the summary of the measurement results **for all measured intersections**.

Table 4. Number of vehicles [veh/312 h] passing through the stop line when displaying the red signal with the countdown timers switched on and off for all measured intersections

	Car	Delivery truck	Minibus	Truck	Bus	Motor and bike
Countdown ON	627	42	22	9	7	1
Countdown OFF	1069	55	22	19	24	1

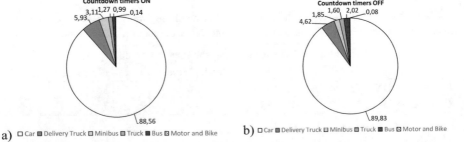

a) □ Car ■ Delivery Truck □ Minibus ▧ Truck ■ Bus ▨ Motor and Bike   b) □ Car ■ Delivery Truck □ Minibus ▧ Truck ■ Bus ▨ Motor and Bike

Fig. 8. The percentage share of particular groups of vehicles passing through the stop line while displaying the red signal with the countdown timers switched on and off for all measured intersections

By analyzing the results presented in Table 4 and in Fig. 8, it can be stated that in the case of switching the countdown timers off, 482 more illegal passing through the stop line have been recorded. Excluding the group of passenger cars, the largest increase in number of vehicles passing through has been noted in the group of delivery trucks and buses (three times more).

In both studied periods the highest percentage share of passes through the stop line in the group of vehicles pertained to the passenger cars (over 88%). Delivery trucks were about 5%. Buses, in turn, accounted for about 2% entries into intersections during the red signal when the countdown timers were switched off and for about 1% - when the devices were on.

In addition, parametric tests for two structure indicators at the significance level of 0.05 were carried out. The following values of the test statistic have been obtained: for particular intersection in Zabrze (-2.797), in Opole (-2.904), in Wrocław (-11.636) and for all measured intersections (-11.151). The results indicate that the differences between the share of vehicles passing through the stop line while displaying the red signal in all cases are statistically significant. It means that these shares depend on the use of the countdown timers.

4   Conclusions

The results of the study show that at each of examined intersection more vehicles passing through the stop line during the red signal have been observed with the countdown timers off than when they were switched on. Therefore, it can be stated that the use of the countdown timers does not contribute to reducing the level of traffic safety - at least in the aspect of the number of vehicles passing through the stop line while the red signal is displayed.

The largest number of the entries into intersections at the red signal has been reported among the passenger car drivers. They were more than 89%, with the average share of passenger cars at 87% in total traffic volume. In other groups of vehicles, the number of passes through the stop line during the red signal was similar, for example for delivery trucks it was 5.11% and the share of these types of vehicles was 6.71% in total traffic volume, while for buses this number was 1.63% with an average share of these vehicles at the level of 1.64%.

The results of the analyzes have confirmed that the countdown timers affect the level of traffic safety **in dependence on the function of crossing streets**. The thesis may be confirmed by the comparison the results for the intersection in Zabrze (with the urban traffic) and for the intersection in Wrocław (with non-urban traffic). At the intersection in Wrocław, the number of vehicles passing through the stop line during the red signal with the countdown timers switched off was 2.2 times higher than when the devices were switched on, and at the intersection in Zabrze, the number was 1.25 times higher. However, it should remember that at the intersection in Wrocław the higher total traffic volumes have been recorded than at the intersection in Zabrze. In addition, the drivers traveling through intersection with non-urban traffic make longer journeys than those traveling within the city. They may be more tired and therefore less focused on correct reading and interpretation of traffic signs and signals. Moreover, not without significance may be the rush of drivers as well as the technical parameters of the linear infrastructure to encourage the driving with the more than safe speed.

A different tendency has been observed among bus drivers. At the intersection in Zabrze there were 5 times more passes through the stop line during the red signal than at the intersection in Wrocław (2.33 times more). Bus drivers decided to break the law, mainly because of the rush and the need to keeping the punctuality.

Accordingly, the results of conducted research confirmed, that introducing the countdown timers at the intersections with non-urban traffic could help eliminate RLR phenomenon and hence increase the level of safety at such junctions.

The research in this field will be continued by the authors of article. They also intend to undertake the research works on influence of the countdown timers on traffic smoothness as well as on the traffic management, what is highlighted in the articles [13, 14].

References

1. Sobota, A., Kłos, M.J., Karoń, G., Żochowska, R.: Modelling of the traffic safety at signalized intersection with countdown timers—methodology of research. (in printing)
2. Kempa, J., Bebyn, G.: Experiences from the operation of countdown timers at traffic signaling devices in Toruń. Logistics 6, 5351–5363 (2014)
3. Kempa, J., Bebyn, G.: Countdown timers at traffic signaling devices. Logistics 6, 5364–5370 (2014)
4. Regulation of the Minister of Infrastructure of July 3, 2003 on detailed technical conditions for road signs and signals as well as traffic safety devices and conditions for their placement on roads. Journal of Laws 220 item 2181 from 2003 (2003)
5. Anjana, S.: Safety analysis of urban signalized intersection under mixed traffic. J. Saf. Res. 52, 9–14 (2015)
6. Karoń, G., Pawlicki, J., Pytel, D.: Analysis of road safety using SEWIK database (accidents and collisions records)—county of Tarnowskie Góry (in quarter 1 of 2008), Contemporary transportation systems. In: Janecki, R., Sierpiński, G. (eds.) Selected Theoretical and Practical Problems. The Development of Transportation Systems. Monograph No. 256, pp. 155–164. Publishing House of Silesian University of Technology, Gliwice (2010)
7. Sobota, A., Tuchowski, J., Żochowska, R.: Analysis of the level of road safety at intersections in Tychy. Logistics 4, 3287–3298 (2014)
8. Szczuraszek, T.: Urban Traffic Safety. Publishing House for Communication, Warsaw (2005)
9. Szczuraszek, T., et al.: Examination of Traffic Hazards. Polish Academy of Sciences, Warsaw (2005)
10. Bonneson, J., Zimmerman, K.: Red-light-running handbook: an engineer's guide to reducing red-light-related crashes. Report No. FHWA/TX-05/0-4196-P1, September 2004
11. Bonneson, J., Brewer, M., Zimmerman, K.: Review and evaluation of factors that affect the frequency of red-light -running. Report No. FHWNTX-02/4027-1, September 2001
12. Wenbo, S., Zhaocheng, H., Xi, X., Feifei, X.: Exploring impacts of countdown timers on queue discharge characteristics of through movement at signalized intersections. Procedia Soc. Behav. Sci. 96, 255–264 (2013)

13. Sobota, A.: Traffic smoothness in the light of research. Contemporary transportation systems. Selected theoretical and practical problems. In: Janecki, R., Sierpiński, G. (eds.) The Development of Transportation Systems. Monograph No. 256, pp. 165–172. Publishing House of Silesian University of Technology, Gliwice (2010)
14. Żochowska, R., Karoń, G., Sobota, A.: Managing traffic congestion in urban networks—selected issues. Logistics **6**(CD-ROM), 11850–11861 (2014)

Reducing Odometer Fraud in the EU Second-Hand Passenger Car Market Through Technical Solution

Przemyslaw Borkowski[✉]

University of Gdansk, Gdansk, Poland
przemyslaw.borkowski@univ.gda.pl

Abstract. One of the key issues while purchasing used car is viability of its odometer reading. Mileage incurred is basis for consumer decisions. European Union second-hand car markets score low in surveys in regard to their honesty and reliability. Evidence from around the world as well as from the EU suggests that odometer tampering is quite frequent practice. Thus dishonest sellers intercept difference in monetary value between odometer tampered and non-tampered car. For buyer there are hidden costs of depreciation and maintenance incurred but not revealed. In this paper the economic viability of one of the possible solutions to fight odometer fraud in regard to passenger cars is discussed. This solution is based on introduction of tamper-proof odometers into the market. The costs and benefits of such a move are estimated and pros and cons of the proposal are discussed. The estimates are produced for different scenarios of assumed odometer fraud levels. Regardless of the fraud level scenario considered the straight monetary benefits outweigh costs. However there are other issues which influence possible introduction of tamper-proof odometers like lasting of their effects or social and technical barriers to their adoption.

Keywords: Odometer tampering
Costs and benefits of tampering-proof odometers
Reducing odometer fraud in the EU

1 Introduction

Odometer fraud occurs whenever an odometer is rolled back in order to purposefully show lower mileage than actually incurred. It results from odometer tampering, also referred to as odometer rollback, "clocking", or odometer manipulation. This is being done in order to improve sale prospects and achieve better selling price of used car.

The magnitude of the phenomenon is rather significant, as per existing evidence, and loses due to odometer tampering are calculated in millions of EUR per year. Odometer tampering is common with older vehicles but is also registered in cases of trading relatively new cars. However it is easier to tamper odometer in older cars as more mileage incurred allows for more room for manipulation. The phenomenon exists all around the world with significant scale being reported from the US [1], Japan [2],

© Springer Nature Switzerland AG 2019
G. Sierpiński (Ed.): TSTP 2018, AISC 844, pp. 184–194, 2019.
https://doi.org/10.1007/978-3-319-99477-2_17

Australia [3]. In the EU it is more visible in EU-13 than EU-15[1] due to the relatively higher share of older cars in EU-13 fleets. In addition odometer fraud is more frequent in imported vehicles than in domestically traded ones. While for cross-border transactions the problem has been analysed in-depth in recent European Parliament study [4], the internal – domestic scale of the odometer tampering in the EU has not been fully researched. This paper is aiming to breach that gap and discuss the scale of odometer tampering in the domestic markets in the EU. The subject of the analysis is domestic passenger car market. Odometer fraud is also present in commercial vehicles market but size of those markets as compared to the personal cars is much smaller with majority of transactions conducted by specialized business entities, thus risk of fraud is lower. In passenger car markets the split between business-to-consumer and consumer-to-consumer transactions is balanced thus a possibility for fraud is higher. Furthermore this research explores specifically the validity of the so called "technical solution" to odometer fraud. Technical solution means that odometers are equipped with specific tamper-proof devices installed by manufacturers which could make odometer tampering extremely difficult from technical point of view.

With old mechanical odometers the whole process was not overly complicated and required only basic mechanical knowledge to perform. Introduction of digital odometers was supposed to make the process more difficult for fraudsters. This is however not the case, digital odometers can be tampered as easily or even easier than old mechanical devices. There are three reported techniques for cracking digital odometer:

- By connecting cable directly to the chip in which odometer data is stored. In this case the dashboard has to be dismantled, odometer unit disassembled and the chip separated, reprogrammed and later reattached to its mount.
- By connecting a serial cable directly to the odometer unit. In this method only dashboard must be disassembled in order to access serial port.
- By connecting car central computer to external diagnostic unit. Since modern cars are equipped with central computer, access to the odometer unit can be acquired by simply connecting to that computer in normal diagnostic mode.

The technology needed to manipulate odometers electronically can be purchased easily and relatively cheaply. Moreover the web search for the term "odometer tampering" produces thousands of hits directing to the appropriate "services". Neither odometer manipulation nor devices used for the purpose are illegal by default. Tampering is usually illegal, except for the repair of broken odometer. The devices themselves are not illegal and are sold as recalibration units for digital dashboards. A range of products are available utilising anything from laptop computers with serial links through to portable handheld units.

---

[1] EU-15 are: Austria, Belgium, Denmark, Finland, France, Germany, Greece, Ireland, Italy, Luxembourg, Portugal, Spain, Sweden, The Netherlands, United Kingdom. EU-13 in this research refers to enlargement countries: Bulgaria, Czech Republic, Estonia, Hungary, Latvia, Lithuania, Poland, Romania, Slovakia, Slovenia, Croatia, Cyprus, Malta.

2 Cost of Odometer Tampering

Regardless of the manipulation method applied the net result is vehicle with altered mileage reading. The total cost of odometer tampering depends on the number of fraudulent transactions among all domestic car trades. Those are composed of business-to-customer transactions whenever car dealer sales car or customer-to-customer transactions whenever the deal is between private persons. Estimating the number of domestic transactions or the size of second-hand car market is a challenge. Data on new car registrations is collected by ACEA [5]. However there is a shortage of statistics in regard to used cars registrations. For the purpose of this study Eurostat data [6] (which supposedly collects all registrations) has been confronted with ACEA new registrations to produce number of used cars registrations. Those numbers are however in some cases not fully reliable and have been confronted either with other national sources and reports to achieve cross-checked figures. The most complete estimates of the proportion between new and used cars traded come from BCA report [7]. The problem with BCA estimates is that they are performed only for selected EU-15 countries. Similar data has been obtained from CIRP estimate for France, Germany, Italy [8]. In addition a very detailed study conducted by Belgian Car-pass (institution dealing with odometer fraud in Belgium) through dealers interviews for selected EU-15 countries (Belgium, Luxembourg, Germany, Netherlands, France) is available [9]. The much more problematic is the situation of EU-13, whereas exact size of second-hand car market is not known. The EU sponsored study [10] which was supposed to look into the case produced only indication that in EU-15 the number of used cars transactions could be as high as 8 times the new car transactions. However recent estimates from Poland by PZPM [11] suggest that the ratio is only slightly above 2:1, while Millward Brown data claims this number to be at 3:1 [12]. It has to be also noted that adoption of factor of 8 as suggested by [10] might produce a number of transactions which is not believable for some smaller EU-15 countries. Besides in the case of EU-13 the very high share of used-cars traded is supplied from cross-trade rather than from internal domestic second-hand market. Accordingly to the "European second-hand car market analysis" imports exceeding 60% were attributed to BG, CY, CZ, EL, LV, MT, PL, RO and SK [13]. The abovementioned study provides and average EU rate of 3.5 as a proportion between used and new cars traded domestically and this rate has been adopted for estimate of domestic used car market in EU-13.

In order to evaluate the total economic cost of odometer tampering the unit cost of one mile rollback is needed as well as estimate of the number of rolled back kilometres per one tampering. The economic cost comes mainly from two sources and both are costs to which buyer is subjected:

- Depreciation cost representing depreciation incurred but not registered.
- Maintenance cost resulting from need to increase maintenance expenditure due to vehicle components being degraded more than registered.

Both costs result from seller interception of price difference between car with tampered odometer and one without odometer tampering. For the buyer on the second-hand market it is mileage of the vehicle which is the most important indicator of the degree to which traded vehicle has been worn out. Thus mileage manipulation leads to distorted depreciation and maintenance perception on the part of the buyer (Table 1).

Table 1. Estimated domestic trade in used passenger cars in the European Union in 2017

No	Country	No of cars	No	Country	No of cars
1	BE	710,525	15	LT	206,688
2	BG	118,331	16	LU	42,220
3	CZ	950,582	17	HU	406,927
4	DK	664,773	18	MT	n/a
5	DE	4,817,765	19	NL	1,616,698
6	EE	89,663	20	AT	918,632
7	IR	459,746	21	PL	1,210,475
8	EL	308,402	22	PT	1,066,243
9	ES	1,234,931	23	RO	367,790
10	FR	4,010,421	24	SI	218,862
11	HR	177,695	25	SK	335,916
12	IT	3,743,944	26	FI	414,851
13	CY	n/a	27	SE	1,441,693
14	LV	58,443	28	UK	7,621,851

Source: Own elaboration based on [7–13]

The question of the size of odometer fraud is more complex. Firstly, there is usually only incidental evidence uncovered during specific police raids or other control activities. Secondly, since odometer fraud is well hidden illegal activity, there is no chance to obtain any regular statistics on the matter. Nevertheless those limitations there are couple of studies out of which the fraud rates could be derived. For instance in Germany, the German Automobil Club Association – ADAC claims that odometer fraud affects as many as 22% of all internal transactions [10]. Yet this number is simple repetition of the number registered during single Munich Police operation against dishonest car dealers. This is therefore incidental evidence. French evidence from the French competition, consumer protection and anti-fraud authority (DGCCRF) reportedly [10] shows a 43% rate of odometer tampered cars. This is an extremely high rate and again has been obtained from incidental evidence. But both cases seem to be supported by car dealers, who in interviews conducted by Car-pass own study claimed that between 30% and 40% of all domestically traded cars were affected with odometer fraud [10]. On the other hand, car-dealers might tend to overestimate the odometer fraud phenomenon in order to prompt authorities to act. In the Swedish market, there are specific CARFAX – a vehicle history reports provider – estimates of odometer tampering, which offer a more comprehensive data set. Accordingly to it odometer fraud ranges from 3.1% to 7% [14]. In the case of Belgium, where anti-odometer fraud actions have been institutionalised with the establishment of Car-pass, the internal rate of tampering has decreased from initial 8.6% to 0.6% [10], thus the 8.6% ratio could be believed as one representing unprotected against odometer manipulation market for EU-15 countries. Additional evidence is acquired from the UK Office of Fair Trading (OFT) study which yielded odometer tampering rates between 5% and 12.5% [15].

For the EU-13 there is significant lack of solid statistical evidence. But for imported cars odometer manipulation could range between 70% and 90% as claimed by dealers active in the market. Internal market rates are usually much lower at half of the internationally traded cars [4] but it still produces as high as 30%–40% rate. Accordingly to Czech member of the European Parliament T. Zdechovsky, Czech Republic Transport Ministry estimated odometer tampered cars in the internal market at 37% [16]. A much lower numbers are produced by EU-wide consumer market study [17] with odometer fraud reported by 1%–7% of consumers in EU-15 second hand car markets and 10%–20% range given by customers buying used cars in EU-13. It has to be however said, that this numbers should be considered a very conservative because the study reports only the uncovered cases of fraud. It is unlikely that customers were able to discover all cases and thus reported rates are lower than real ones.

For the purpose of this research two odometer manipulation rates for two scenarios: lower and higher estimate are adopted. The lower estimate scenario applies numbers from the EU consumer market study [17] and constitutes lower boundary of the estimate as certainly odometer fraud is no less than this number. Second scenario is based on average numbers produced by various evidence as quoted above with the exception of Scandinavia whereas a very precise evidence is available [14], UK and Ireland [15] where also precise estimate is available, and Belgium and Netherlands where anti-tampering institutional solutions are in effect. For the remaining EU-15, 20% average rate is used while for EU-13 – 30%.

Finally the question about the average rollback per one tampering attempt remains. The existing evidence points at the typical range between 30-90 thousands vkms (vehicle-kilometres) rolled back. ADAC report from 2005 produces 33000 km [10], UK evidence gives as much as 108811 km [15]. The annual Car-pass reports give average roll-backs between 45000 and 90000 vkms [18]. For the purpose of this study two variants are adopted: 30000 and 60000 vkms of average rollback. Obviously there is an incidental evidence which reports rollbacks of even hundred thousand vkms. However one has to remember that for odometer fraud to be viable it has to fall within some reasonable range. 30 thousand vkms especially for newer vehicle markets in EU - 15 seems to be plausible, while 60 thousand is more applicable to EU-13 markets.

The final item necessary in order to calculate monetary loss caused by odometer fraud is average depreciation and maintenance cost per one missing vkm. In this study the evidence from DG TREN study comparing depreciation and maintenance costs in different EU countries is used [19]. As a basis for calculation German unit costs are accepted with 0.16 eurocent as depreciation cost of one vkm and 0.05 as maintenance cost of one vkm. Those values are recalculated for other countries by factoring in automotive market average price difference estimated from automotive national markets comparison [20]. The adoption of German market unit costs as a basis is dictated by the fact that German data bases on long time-series and also data collection methodology supports its reliability. Furthermore Germany is the biggest car exporter in the EU thus further internal market trades are dominated (especially in EU-13) by cars produced in Germany. Table 2 summarizes estimated monetary cost of odometer tampering for EU countries under low and high assumed manipulation rates and under 30 and 60 thousands vkms of assumed tampering average rollbacks.

Table 2. Estimated cost of odometer tampering in domestic passenger cars trade in the European Union in 2017 [EUR million]

No	Country	Low tampering - low rate	Low tampering- medium rate	Medium tampering-low rate	Medium tampering- medium rate
1	BE	24.34	24.34	48.67	48.67
2	BG	121.52	182.28	243.04	364.57
3	CZ	402.34	1,508.79	804.69	3,017.58
4	DK	397.45	662.42	794.90	1,324.84
5	DE	607.04	6,070.38	1,214.08	12,140.77
6	EE	40.18	150.68	80.36	301.36
7	IR	30.75	153.76	61.51	307.53
8	EL	160.21	356.02	320.42	712.04
9	ES	704.26	1,408.51	1,408.51	2,817.03
10	FR	508.77	5,087.69	1,017.54	10,175.38
11	HR	65.85	219.50	131.70	439.00
12	IT	1,513.49	4,324.27	3,026.99	8,648.53
13	CY	n/a	n/a	n/a	n/a
14	LV	30.95	103.15	61.89	206.31
15	LT	109.44	364.81	218.89	729.62
16	LU	6.00	24.02	12.01	48.03
17	HU	310.34	477.44	620.67	954.88
18	MT	n/a	n/a	n/a	n/a
19	NL	317.36	528.94	634.73	1,057.88
20	AT	113.50	1,135.02	227.00	2,270.04
21	PL	957.66	1,915.33	1,915.33	3,830.65
22	PT	312.45	1,562.26	624.90	3,124.51
23	RO	302.17	566.56	604.33	1,133.12
24	SL	135.18	405.53	270.35	811.06
25	SK	177.17	531.52	354.34	1,063.03
26	FI	101.75	237.42	203.50	474.84
27	SE	260.66	588.58	521.32	1,177.17
28	UK	1,204.84	2,008.06	2,409.67	4,016.12
29	Total	8,667.00	29,056.74	17,334.00	58,113.47

Source: Own estimates

Even for low fraud rates scenario this is staggering amount of more than 8 billion EUR, while for higher fraud rate scenario this number could reach as much as 58 billion EUR.

3   Technical Measures in Reducing Odometer Fraud in the EU

There are three methods which can be used in order to prevent odometer fraud:

- Odometers readings register.
- Technical solution.
- More frequent control of vehicle vendors.

In this paper the pros and cons of technical solution are addressed in depth. The technical solution requires that odometer is equipped with tamper-proof mechanism. This type of protection requires additional cryptographic keys and data chip which is protected against cracking. Interestingly this type of protection is already used in vehicle on-board anti-theft system. The existing technology could be to high degree reused for the purpose of odometer data protection. Moreover it is reportedly a solution requiring minimal additional financial expenditure, an additional unit cost of adding anti-tamper protection could be as low as 1 EUR [21]. The total cost of implementing technical solution is calculated in Table 3.

Table 3. Estimated cost of equipping passenger cars with tamper-proof odometers in the European Union in 2017 [EUR million]

No	Country	New pass. cars	All fleet	No	Country	New pass. cars	All fleet
1	BE	0.55	446.99	15	LT	0.03	99.53
2	BG	0.03	191.10	16	LU	0.05	30.49
3	CZ	0.27	409.23	17	HU	0.12	255.37
4	DK	0.22	192.33	18	MT	n/a	21.28
5	DE	3.44	3,605.70	19	NL	0.41	666.91
6	EE	0.03	54.13	20	AT	0.35	379.84
7	IR	0.13	158.81	21	PL	0.48	1,657.87
8	EL	0.09	408.39	22	PT	0.22	363.04
9	ES	1.23	1,788.44	23	RO	0.11	412.25
10	FR	2.11	2,553.24	24	SI	0.06	90.47
11	HR	0.05	119.15	25	SK	0.10	163.02
12	IT	1.97	2,988.10	26	FI	0.12	209.03
13	CY	n/a	38.28	27	SE	0.38	373.53
14	LV	0.02	54.20	28	UK	2.54	2,683.40
					Total	14,698,883	19,884.25

Source: Own estimates

The calculation of total cost is made on the basis of new registrations provided by ACEA [5] and supplemented by Eurostat for Bulgaria [6] times the cost of 1 EUR. The calculation assumes that only the odometer unit needs to be protected and other car units which also store mileage information are left without anti-tampering devices. This is very cheap solution, if only for cost consideration, should be compulsory in all newly manufactured vehicles. The yearly cost as per number of newly registered passenger cars in the EU should be as little as 14.7 million EUR, easily bearable by either manufacturers or customers. Yet, this move will not eliminate the problem with older vehicles. Unless odometers are replaced in all existing vehicles the odometer fraud problem could only be eliminated in a very long perspective with older cars dying out of natural scrappage processes. Given the increasing average age of vehicles in the EU, it might result in full eradication of fraud only after 20–25 years. The alternative is immediate recall of all vehicles to manufacturer's workshops in order to replace odometers with new ones – tamper proof. Considering current odometer prices and workload necessary to replace them, the unit cost of such an operation should be – depending on car make and market - between 50 and 400 EUR. For the purpose of this study, the cost is assumed to be 80 EUR including the cost of parts and the necessary workload (as a reference estimated cost of tachograph replacement procedure is adopted [22]). The labour cost difference among Member States as provided by Eurostat is also taken into consideration. Replacement is very car make-depended and might require less or more time. This is, however, one time investment.

The expected benefits form the application of technical solution come from reduction of unaccounted for user depreciation and maintenance costs. The exact scale of those benefits depends directly on the efficiency of the anti-tamper mechanism. If an anti-tampering mechanism is indeed fully tamper-proof than fraud is eliminated completely. But it is likely that, as with any technical solution, there are counter-measures which might be applied by crooks. For the purpose of this study the efficiency of protection mechanism applicable in protecting credit cards is applied. This is mainly because chip-protection to be applied in tamper-proof odometers are similar to those applicable to credit cards. Deployment of technology standard based on chip protection in the credit cards resulted in counterfeit fraud reduction on the scale of 70% [23]. The benefits resulting from reduction of fraud in odometer manipulation applying similar rate are given in Table 4.

Under all variants the potential benefits from deploying technical solution outweigh the cost of this solution. Thus the conclusion seems obvious – it is worth considering and implementing into EU passenger car market. The only remaining question is that about longevity of this solution.

Table 4. Estimated benefits of introducing tamper-proof odometers in passenger cars domestic trade in the European Union [EUR million]

No	Country	Low tampering - low rate	Low tampering- medium rate	Medium tampering-low rate	Medium tampering- medium rate
1	BE	17.03	17.03	34.07	34.07
2	BG	85.07	127.60	170.13	255.20
3	CZ	281.64	1,056.15	563.28	2,112.31
4	DK	278.22	463.69	556.43	927.39
5	DE	424.93	4,249.27	849.85	8,498.54
6	EE	28.13	105.48	56.25	210.95
7	IR	21.53	107.63	43.05	215.27
8	EL	112.15	249.21	224.29	498.43
9	ES	492.98	985.96	985.96	1,971.92
10	FR	356.14	3,561.38	712.28	7,122.76
11	HR	46.10	153.65	92.19	307.30
12	IT	1,059.45	3,026.99	2,118.89	6,053.97
13	CY	n/a	n/a	n/a	n/a
14	LV	21.66	72.21	43.32	144.42
15	LT	76.61	255.37	153.22	510.74
16	LU	4.20	16.81	8.41	33.62
17	HU	217.24	334.21	434.47	668.42
18	MT	n/a	n/a	n/a	n/a
19	NL	222.16	370.26	444.31	740.52
20	AT	79.45	794.52	158.90	1,589.03
21	PL	670.36	1,340.73	1,340.73	2,681.46
22	PT	218.72	1,093.58	437.43	2,187.16
23	RO	211.52	396.59	423.03	793.18
24	SL	94.62	283.87	189.25	567.75
25	SK	124.02	372.06	248.04	744.12
26	FI	71.23	166.19	142.45	332.39
27	SE	182.46	412.01	364.92	824.02
28	UK	843.39	1,405.64	1,686.77	2,811.29
29	Total	6,066.90	20,339.72	12,133.80	40,679.43

Source: Own estimates

4   Conclusions and Discussion

Discussing efficiency of the proposed technical solution to the odometer fraud problem, firstly the decision whether to apply new tamper-proof odometers to only newly produced vehicles or to the whole existing fleet of vehicles has to be addressed. If use of tamper-proof devices is limited only to newly produced vehicles, this is somewhat contradictory to the goal of putting an end to the tampering in a short time. The result will be rather prolonged process in which tamper-proof and no tamper-proof odometer vehicles will coexist on the market for long years. One has also to remember that odometers tend to be tampered more frequently in older vehicles. Therefore it should be expected that rate of odometer tampering will seriously fall only when newly produced tamper-proof odometer vehicles constitute majority of second-hand traded vehicles. This might take close to a decade and in some countries where older vehicles constitute bigger share of the whole vehicle fleet (most of EU-13) even longer. But the one time replacement comes at a very significant cost. Should manufacturers or car owners be responsible for paying this costs? Besides it is doubtful that there are sufficient resources (new tamper-proof odometer units as well as sufficient number of certified workshops) to perform such a replacement in relatively short time.

Secondly, the calculation presented assumes that only the odometer unit needs to be protected and other car components, some of which also store mileage information, are unprotected. With perfectly working main odometer protection there is no need for any safeguards in other car components. But if main odometer unit is cracked additional protected information storage could be valuable. There is also a question about cryptographic solution to be used. International Automobile Federation (FIA) points out [21] that since the United Nations Economic Commission for Europe (UNECE) Regulation 18 on anti-theft of motor vehicles, systems such as Secure Hardware Extension (SHE) or Hardware Secure Module (HSM) are already fitted in vehicles. Manufacturers already have access to cryptographic keys that block unauthorised access, which is why it is relatively cheap to implement protection into the odometer unit. Another possible reuse is that of software which provides protection in anti-theft devices and is as well already installed in the car. It is a simple technical task to reengineer it to apply to odometers.

Thirdly, the application of additional protection makes it more difficult for honest users to replace damaged odometer. Certainly damage or malfunction of the odometer are rare occurrences. But if it happens, only specialised and probably only official factory workshops would be able to replace odometer. Non-factory workshops are unlikely to be given cryptographic keys to do so because if they are passed to all workshops it would make cracking their protection easy, which in turn beats the purpose of installing advanced odometer safeguards.

Finally although the presented technical solution is tempting due to its operational simplicity and relatively low cost (at least under the scenario of application to only newly produced cars), the main problem with it, apart from the replacement cost, is its questionable permanence. Any technical solution based on software can be cracked. And if cracked should it be replaced again by more advanced protection? Cost and servicing issues are the main factors here. Summarising, technical anti-tampering protection of odometers seems viable option but to achieve full efficiency should be rather supplemented by other - like e.g. mileage registers -means.

References

1. Kelly, J.: A Study and Statistical Analysis of Odometer Tampering by Purchasers of One-Time Lease Vehicles. NHTSA, Washington (1992)
2. http://www.japaneseodometercheck.com/
3. South Australia Police: https://www.cbs.sa.gov.au/assets/files/numbers_up_for_odometer_clocking.pdf
4. European Parliament: Odometer Manipulation in Motor Vehicles in the EU. http://www.europarl.europa.eu/RegData/etudes/STUD/2018/615637/EPRS_STU(2018)615637_EN.pdf
5. European Automobile Manufacturers' Association: ACEA Report Vehicles in use Europe 2017.   http://www.acea.be/uploads/statistic_documents/ACEA_Report_Vehicles_in_use-Europe_2017.pdf
6. Eurostat: http://ec.europa.eu/eurostat/web/transport/data/database
7. University of Buckingham: The used car market report 2014. A Report for BCA. Centre for Automotive Management, Birmingham (2014)
8. CIRP II, Capgemini: The Anatomy and Physiology of the Used Car Business. https://www.capgemini.com/wp-content/uploads/2017/07/tl_Anatomy_and_Physiology_of_the_Used_Car_Business.pdf
9. Venherle, K., Vergeer, R.: Data gathering and analysis to improve the understanding of 2nd hand car and LDV markets and implications for the cost effectiveness and social equity of LDV CO2 regulations, Final Report for: DG Climate Action, Brussels (2016)
10. Car-Pass: Impact Study of Mileage Fraud with Used Cars & Adaptability of the Car-Pass Model in Other EU-Countries. Brussels (2010)
11. Polish Association of Motor Industry. http://www.pzpm.org.pl/en/Media/Files/Raport-PZPM-2016
12. Millward Brown: Autowybory Polaków (2016)
13. Mehlhart, G., Merz, C., Akkermans, L., Jordal-Jorgensen, J.: European Second-Hand Car Market Analysis, Final Report. Darmstadt (2011)
14. Carfax: https://www.carfax.eu/article/odometer-fraud-europe.html
15. Office of Fair Traiding: The Second-Hand Car Market. An OFT Market Study, London (2010)
16. Euractive: https://www.euractiv.com/section/road-safety/opinion/odometer-fraud-resonates-across-the-whole-european-union/#comment-326374
17. European Commission: http://ec.europa.eu/consumers/consumer_evidence/market_studies/docs/2ndhandcarsreportpart1_synthesisreport_en.pdf
18. Car-Pass Website: https://www.car-pass.be/en/about-car-pass
19. Maibach, M., Peter, M., Sutter, D.: Analysis of operating cost in the EU and the US. Annex 1 to Final Report of COMPETE Analysis of the contribution of transport policies to the competitiveness of the EU economy and comparison with the United States. Funded by European Commission. DG TREN, Karlsruhe (2006)
20. Carspring: Global Used Car Price Index (2017)
21. Fédération Internationale de l'Automobile: http://www.fiaregion1.com/wp-content/uploads/2014/04/2017-07-04-Mileage-Fraud-with-new-logo.pdf
22. Suchanek, M.: Retrofitting Smart Tachographs by 2020. In-depth analysis for the European Parliament. Brussels (2018)
23. Barclays: Barclays' Security in Payments: A Look into Fraud, Fraud Prevention, & the Future (2015)

Integrated Urban Passenger and Freight Transport

Integration of Public Transport Modes – Case Study of Wieliczka, Krakow Agglomeration

Aleksandra Ciaston-Ciulkin and Sabina Pulawska-Obiedowska$^{(\boxtimes)}$

Faculty of Civil Engineering, Cracow University of Technology,
Krakow, Poland
{aciaston-ciulkin, spulawska}@pk.edu.pl

Abstract. Increasing problems resulting from the growing mobility and travel elongation of the suburban area inhabitants force to search for solutions that would help sustainable transport development. The process of sustainable development supports integration of various transport modes. The case study of the integration of public transport modes operating within the Krakow agglomeration area can be presented as an example of a good practice for integration of rail with bus systems. The article presents solutions applied in the field of railway connection integration with bus lines. It also presents the degree of the use of bus lines in integration with the railway connection.

Keywords: Agglomeration railway · Krakow · Transport system integration
Rail and bus integration · Seamless transport · Public transport integration

1 Introduction

The integration of transport systems is a subject of scientific research [e.g. 1–6], an instrument of European transport policy [e.g. 7–12], and a method of practical solutions in transport [e.g. 13–16]. Integration plays a large role in both passenger and freight transport. The article focuses on the issues of passenger transport integration. The concept of transport integration is common and has been lately more and more precisely defined. Although multiple interpretations of this concept are still possible, the broad definition of transport integration understands it as a process of making available all possible transport modes to all users, which results in the most effective, efficient, and smooth way of movement [17–19]. Perfect integration is a condition for 'seamless transport' [20].

Achieving the overarching goal of integration allows to treat it as a certain process of transport transformation. In this sense, the most accurate definition seems to be the one proposed in 2003 by the research team NEA and the University of Oxford [4]. According to the authors, integration is an organized process in which the technical components of the passenger transport system (network and infrastructure), tariffs and ticketing systems, together with the information and marketing of various operators using different modes of transport, interact and work better and more efficiently. The result of this cooperation is a general improvement in the quality of services combined also with elements of individual transport modes.

G. Sierpiński (Ed.): TSTP 2018, AISC 844, pp. 197–207, 2019.
https://doi.org/10.1007/978-3-319-99477-2_18

Transport integration plays an important role in urban areas struggling with congestion, environmental pollution, and noise. The integration is a source of many benefits, both individual and social. The integration in passenger transport offers the possibility of promoting public transport, increasing transport accessibility, ensuring smooth travelling, increasing the quality of transport services, reducing transport costs, and minimizing negative impact on the environment.

The opportunity to maximize the benefits of passenger transport integration is associated with the merging of various complementary integration modules [6, 21, 22]. The most important ones are the infrastructural, informational, organizational, economic/financial, and spatial modules.

Infrastructure integration is aimed to merge the technical elements of the system, like stops, stations, infrastructure connections for the passenger, taking into account changes between not only public types of transport, but also individual ones. It should be taken into consideration that in passenger transport, pedestrian movement is treated as one of the transport modes, which is why the infrastructure integration process involves building, improving, and encouraging pedestrian movement infrastructure. Therefore, in infrastructure integration, aside from common transport networks, shared stops, tram and bus common tracks, interchange nodes, Park & Ride and Bike & Ride parkings, elements facilitating and encouraging pedestrian traffic, such as footbridges, pedestrian routes, access points, lifts, etc. are equally important. Infrastructure integration solutions contribute not only to shorter transfer times and convenience of travelling, but also to an increase in travelling safety, as it reduces the contact of pedestrians with individual vehicles.

The second important issue is the organizational integration consisting in unifying the frequency of particular transport modes and synchronization of timetables in order to ensure the continuity of travel and minimize the transfer time. The coordination of timetables should consider a comfortable change time for passengers and the appropriate arrangement of courses during the day, tailored to the passenger's needs. The information accessibility of a current situation is also becoming more and more important.

A crucial manifestation of the integration is the unification of the economic and financial conditions of travel regardless of the type of operator or public transport mode. Unification of the tariff system, uniformed fees and integrated tickets are introduced. Often, tariff integration is also accompanied by the local integration of line schemes, regulations, tariff zone boundaries, ticket style, regulation on reduced or additional fees.

Complementary to the organizational and economic integration is an integrated information system, which means that travel information can be found in a common platform. Integration in the field of information system concerns both information regarding orientation in the field (all kinds of signs during the trip), as well as information on a specific journey (route selection, location of stops, interchanges, number of transfers, travel time, travel cost, occurring difficulties, etc.). Individual information can be obtained at stops (timetable, charges, tariff), inside the vehicle (additional fees, tariffs, regulations, current route), at information points, by phone, and on the Internet (travel planners and mobile applications).

Another important module is the spatial dimension of integration which refers to urban planning linked to the existing or planned transport network. Spatial planning is correlated with the development of urbanized areas. At the same time, planning the development of the transport system is a significant element in spatial planning. Therefore, it is important that spatial planning and transport development are integrated processes.

One may point out more and more examples of good practices that show how to integrate transport modes and processes to perform high-quality passenger transport services 'from door to door' and do it seamlessly. One of such examples is the introduction of the shuttle bus lines operating to the railway station located in Wieliczka, a small town located in the Krakow agglomeration.

2 Shuttle Bus Lines for Passengers Using the Railway Connection Krakow - Wieliczka

Krakow and Wieliczka are cities located in the south-eastern part of Poland. Kraków, as the capital of the region, is the administrative, cultural, educational, scientific, economic, service and tourist center. It has over 765 thousand residents. Wieliczka, on the other hand, is an urban-rural commune adjacent to Krakow from the south-east side. It is part of the Krakow agglomeration. The commune is inhabited by approx. 52 thousand people. The capital of the commune is the city of Wieliczka, located about 12 km from Krakow. The population of Wieliczka city counts about 22 thousand inhabitants and is a significant economic and educational center for other residents of the commune, but also an important tourist center in southern Poland (the Salt Mine in Wieliczka is inscribed on the UNESCO list). The commune of Wieliczka itself is often called 'the bedroom of Krakow', which is associated with the generation of a large number of daily trips from villages in the commune of Wieliczka to Krakow.

The large number of daily trips affects the need for efficient transport system in this area. People travelling in particular between the cities of Wieliczka and Krakow have a rich transport offer at their disposal. In addition to individual transport, available services are:

- bus public transport - 4 agglomeration lines performing a total of 8–10 courses per hour;
- private collective transport - about 6–8 courses in an hour;
- train - 2 courses in an hour.

Although the railway offer has the lowest frequency of connections, it competes efficiently with other offers in the scopes of [22, 23]:

- immutability of the travel time (almost twice shorter than the travel time by other means of transport and independent of road conditions),
- regularity and punctuality,
- high comfort of travelling,
- lower transport costs compared to other means of transport.

For people living in the surroundings of the Wieliczka city, the railway offer is unfortunately less competitive due to the need to reach the starting station. In order to expand the area of impact of the railway connection, many solutions were implemented to integrate the railway connection with other transport modes. Among them one may indicate Park & Ride car parks, Bike & Ride, integrated stops, or a common transport tariff.

A good solution for transport integration in the area of the Wieliczka commune is the provision of bus lines dedicated for train passengers to/from the final railway station in Wieliczka. The main objective of their activation is to increase the area of influence of the railway line and to facilitate the residents of Wieliczka commune to reach the railway station. Bus lines were established in December 2014 in addition to the change of the railway operator and the improvement of the railway transport offer. Initially, there were two lines: B1 on the Wieliczka - Byszyce route and R1 on the Wieliczka - Raciborsko route. As a result of a big interest in the bus and train offer, shortly after, the R1 line was extended to Grajów, and another line has been launched: line D1 from Wieliczka to Dobranowice (and then to Huciska). Currently (from 1 September 2017), there exist 6 bus lines serving the Wieliczka Rynek Kopalnia railway station. The previous bus offer has been extended by three more lines: S1 (on the route Wieliczka - Świątniki Górne), G1 (on the Wieliczka - Grabie route) and G2 (being a combination of the G1 and S1 lines) – see table below (Table 1).

Table 1. Characteristics of shuttle buses routes.

Line	B1	D1	R1	G1	S1	G2
Route	Wieliczka – Byszyce	Wieliczka – Hucisko	Wieliczka – Grajów	Wieliczka – Węgrzce Wielkie	Wieliczka – Świątniki	Świątniki – Węgrzce Wielkie
Number of stops	24	22	22	17	28	44
The route length	12 km	9 km	8 km	7 km	12 km	19
Number of courses	18 courses in each direction on a working day (each line serves half of the arrivals and departures of trains during the day); on Saturdays, Sundays and holidays, the frequency is twice lower.					2 courses in each direction

The last node of the connections is a bus loop located in Wieliczka called "Wieliczka Rynek Kopalnia". The bus loop is located directly next to the railway station, which enables convenient transfer and waiting conditions for vehicles of both transport modes. It also affects the reduction of transfer time, increases safety and comfort of passengers during transfer. Stops are equipped with electronic timetables, city maps and pavements, convenient also for people with restricted mobility.

Timetables for trains and buses are coordinated. The time between the train arrival and the bus departure, and between the bus arrival and the train departure is about 10 min. It is sufficient for free movement between vehicles, and, at the same time, does not discourage passengers with excessive waiting time. The attractiveness of the bus and train connections is additionally increased by the tariff integration. Passengers purchase one ticket valid both on the train and in the bus. The ticket offer includes single or season tickets.

The former option are 2-h tickets (regular or reduced) or 4-h tickets (for seniors). Single tickets are valid only to the borders of the Wieliczka poviat (to the Wieliczka Bogucice railway stop). Further travel by train requires a purchase of a separate ticket. The cost of a single regular 2-h ticket is 3.30 PLN. Monthly tickets are valid for one month both in buses and on trains. The cost of a regular ticket is 80 PLN or 160 PLN. The cheaper ticket entitles the passenger to take the bus and train to the city boundaries only. The second one is valid on bus routes and on the train up to the Kraków Olszanica station. Different reduced fees are included in the tariff.

3   The Importance of Bus and Rail Integration in Shaping the Demand for Rail Transport

3.1   Number of Integrated Passenger Journeys

The development of bus lines to the railway station in Wieliczka, in particular, increases the spatial accessibility of the railway offer and also increases the role of public transport in everyday trips. The opening of bus connections also increases the accessibility of the city of Wieliczka. Not all passengers using bus lines B1, D1 and R1 also use the railway connection to Krakow. The final destination for the part of bus passengers is the city of Wieliczka.

In order to determine the significance of bus and train integration after three years of its operation, a research was conducted on the number of passengers using this solution. The study has been carried out in February 2017, when the number of people travelling by bus to/from Wieliczka, including people transferring to/from trains, has been counted [24]. Additionally, based on data on the total number of people using the Wieliczka Rynek Kopalnia railway station [25], the share of integrated journeys at this station was determined. The observation was carried out only on three bus lines B1, R1 and D1, because the other bus lines were launched only in September 2017.

The B1, D1 and R1 lines are used daily by nearly 750 passengers. The number of passengers using bus lines in specific directions is balanced. The share of particular bus lines in serving the demand is relatively the same. The least number of passengers use the D1 bus route (28%). Bus numbers B1 and R1 are used by the same number of passengers (market share of 36%).

Bus lines B1, D1 and R1 have a high share in servicing passengers using a railway connection. This share is uneven in both directions and depending on the part of the day. In relation to Wieliczka, approx. 62% of all passengers travelling by bus lines transfer to the train. In the reverse relation, around 55% of bus line passengers change from the train. Differences in both directions result mainly from the fact that in the morning passengers who arrive by bus to Wieliczka directly change to the train due to undertaking obligatory trips to Krakow (school, work). In return trips (from Krakow), some passengers get off the train and carry out additional destinations in Wieliczka (e.g. shopping, official matters, doctor, family, etc.), and then return home by bus.

The average number of passengers in the bus line course is approx. 6.9. B1 and R1 lines are characterized by a similar average number of passengers per course (7.4), while the D1 line has the smallest average number of passengers per course (6). This number is usually exceeded in the courses provided in the morning hours to Wieliczka and in the courses realized in the afternoon hours from Wieliczka.

In thc morning, between 5:00 and 8:59, one may clearly observe that a high number of passengers in the buses to Wieliczka is apparently maintained. Most passengers arrive between 7:00 and 8:59 (30% of all passengers). Such a high result is caused by the obligatory trips of the commune inhabitants to schools and jobs. The final destination of this part of trips is mainly located in Krakow - over 90% of them change to a train and continue their journey. Less than 10% of bus passengers end their journey in Wieliczka. In the afternoon, the importance of bus lines in servicing the railway connection is decreasing, and the importance of commuting to Wieliczka is growing. After 10:00 am, at least half of bus line passengers end their journey in Wieliczka and do not change to the train (Fig. 1).

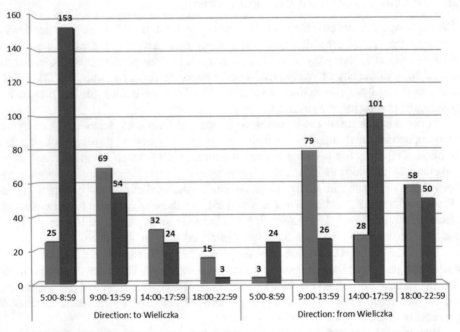

Fig. 1. The number of passengers of shuttle bus lines B1, D1 and R1 to/from Wieliczka (Source: own elaboration)

A different situation can be observed in the opposite direction. The least number of passengers (about 5%) travel by lines B1, D1 and R1 from Wieliczka up to 8:00 am. However, almost all passengers of bus lines in the early morning transfer from the train. It is usually associated with commuting from work in the night shift. Most passengers use bus connections in the afternoon. However, it should be noted that the afternoon peak is much longer than the morning peak. The largest passenger flows are observed from 14:00 to 18:59. At that time, 45% of all passengers use buses from Wieliczka to neighbouring villages.

However, it is worth paying attention to the varied degree of bus and rail integration during these hours. Around a quarter of all passengers travel between 14:00 and 16:59. About 90% of them transfer from the train. Between 17:00–18:59 one may observe the biggest bus passenger flow, but only about 42% of passengers continue the journey started by train. The rest of passengers starts their travel in the city of Wieliczka. However, it should be noticed that such a situation is not related only to the greater role of bus connections in the handling of journeys starting in Wieliczka. In this group of passengers are also those who have previously arrived from Krakow by train, realized other destinations in Wieliczka, and then continue their journey by bus. The adopted methodology of the study did not allow to determine the size of this passenger group in detail.

3.2   The Share of Integrated Journeys in the Size of Passenger Flows at the Wieliczka Rynek Kopalnia Railway Station

The Wieliczka Rynek Kopalnia railway station is used by over 2.9 thousand passengers in a day, of which nearly 1.6 thousand travel in the direction to Krakow and over 1.3 thousand travel in the opposite direction. The difference in the flow size is most probably caused by the need of reaching - in addition to obligatory trip purpose - additional destination during the way back home and subsequent selection of other transport modes, such as agglomeration buses or carshared journeys, which are also convenient for access to Wieliczka.

The distribution of the passenger number is diversified during the day. The peak hours are clearly outlined. The biggest passenger flow may be observed during the morning peak in the direction to Krakow and during the afternoon peak - from Krakow (about 40% of all passengers). Almost 450 rail passengers to or from the Wieliczka station arrive or depart by shuttle bus lines. In the total number of passengers using the analyzed train stop has a 15% share. This means that on average every seventh train passenger uses the integration of rail and bus connections per day.

Most passengers use the integration of both systems travel to Krakow during the morning peak hours. By 9:00 almost every fourth passenger departing by train has previously traveled on one of three shuttle bus lines. The share of passengers using the integrated journey in relation to all passengers departing from the Wieliczka Rynek Kopalnia station after 9:00 am is decreasing:

- 15.2% between 9:00 and 13:59,
- 5.4% between 14:00 and 17:59,
- 3% between 18:00 and 22:59.

In the opposite direction, the usage of the integration solution is the biggest during the afternoon. From 3 pm, on average, every fifth passenger getting off the train in Wieliczka switches to one of the shuttle bus lines. In the morning and early afternoon, the usage of the integration is around 8%. The maximum values of the share of integrated trips in the total number of rail trips at the Wieliczka Rynek Kopalnia station are:

- 52% in the direction to Krakow (9:00 am–9:59)
- 50% in the direction from Krakow (21:00–21:59).

Clearly outlined communication peaks indicate a significant use of the shuttle bus lines in obligatory everyday trips (work, school). Particularly clear peaks are visible in the case of integrated trips (bus and train), which underlines the role and high importance of the integration solution in shaping the demand for public transport. Increasing the accessibility of the railway offer beyond the area marked out with its infrastructure results in:

- nearly 450 people using an integrated transport system during the working days; annually, it is over 100,000 passengers.
- an increase in the passenger flow at the Wieliczka Rynek Kopalnia railway station by 17.5%.

4   Summary

The increasing mobility of inhabitants, especially in urban areas, has consequences related to road congestion and air pollution. An uncontrolled increase in the number of trips usually leads to excessive use of individual transport and the intensification of the negative effects of transport. That is why actions aimed at sustainable development of transport are extremely important, which in urban areas might be manifested in the promotion of integrated journeys under various means of transport. The integration of various transport modes serves to reduce the role of individual transport while maintaining the smoothest traffic flow. The pursuit of 'seamless transport' takes place using a wide range of tools, in particular in the field of infrastructural, organizational, economic and financial, informational, and spatial integration.

A good example of integration in the field of passenger transport is the solution introduced in Wieliczka, the town within the Krakow agglomeration. The development of shuttle bus lines to service passengers to the train from the villages neighbouring Wieliczka has been set. In order to keep a smooth flow and minimize the negative effects of transfers from buses to train, a number of solutions integrating two systems has been applied:

- the final bus stop is located in the immediate vicinity of the railway station (about 10 m);
- the bus timetable is synchronized with the train schedule;
- on the bus lines and on the train, a common transport tariff is applied (periodic tickets are fully integrated, single tickets are partially integrated);
- information on bus and train connections is available on the railway operator's website and at the interchange hub in Wieliczka.

The effectiveness and legitimacy of the implemented solutions is evidenced by the interest in the bus lines among passengers. About 750 passengers per day use the shuttle bus lines, of which 60% travel on an integrated journey. The launching of the bus lines increased the accessibility of the railway connection linking Wieliczka and Krakow, and, at the same time, minimized commuting to the Wieliczka station by means of individual transport. Bus lines users do not have to worry about parking space, nor are they interested in parking fees. On average, one bus (6.9 people/course) replaces four cars (from 1.6 to 2.2 people/vehicle depending on the travel motivation) [26]. A big advantage of the implemented solutions is the preservation of competitive travel time by minimizing the time needed for interchange, as well as the convenience of travel, and the improvement of safety at the interchange point. All these advantages make bus lines have a significant share in passenger service using the Wieliczka Rynek Kopalnia railway station. Passengers switching from bus to train or vice versa constitute on average 15% of all users of the railway station, and in some trains, even one half.

Intelligent tools play a huge role in the success of the introduced solution: an integrated ticket tariff gives passengers the advantage of using only one ticket, a coordinated timetable minimalizes the time lost for a change, and all these benefits are further intensified by an integrated transfer node. The illustrated example of integration is a response to individual expectations regarding access to public transport services and social expectations regarding sustainable transport development. The benefits of this type of solution, as well as the interest among passengers, which were indicated in the article, are a recommendation for the wider use of shuttle bus lines not only in the Krakow agglomeration, but also for any other city or area that is developing the agglomeration rail system.

Such a system of bus lines is justified when developed for the service of final or intermediate stations located in small and medium urban centers with a significant suburbanisation level, as in the described case study. The introduction of shuttle bus lines in such a case results in a specific extension of the railway line and an increase in its asccessibility. In the case of the Krakow agglomeration, such a solution could bring effects similar to those in Wieliczka, e.g. in Sędziszów, Miechów or Skawina (SKA2 line), or in Bochnia, Brzesko, Krzyszkowice or in Trzebinia (SKA3 line).

In the case of the agglomeration rail system, it is also advisable to increase its accessibility by introducing shuttle bus lines to small intermediate stations located on a given line. The legitimacy of such a solution is particular in the case when the railway line runs far from large traffic generators. In the Krakow agglomeration, this is the case, for example, in the city of Niepołomice and neighboring villages. They are located beyond direct access to the railway line, and generate large streams of daily trips to and from Krakow. Access to the railway offer would be provided by shuttle bus lines servicing, for example, the Staniątki station.

The development of shuttle bus lines can also be considered in cities that have access to the agglomeration railways, but the railway station is located in peripheral area and is out of reach by walking. The organization of shuttle bus lines can be even more important in cities that do not have their own public transport, thus make passengers travel to the railway station by car or just give up a railway connection. A good example of such a situation is the city of Miechów served by the SKA2 line. The railway station in Miechów is located on the eastern border of the city, which means that for over half of the residents it is more than 2 km away.

The agglomeration rail system systematically supported by the development of shuttle bus lines in Małopolska can compete more effectively with individual transport modes. However, the development of such a solution needs the highest possible level of integration of train-bus systems, and this requires close cooperation of the railway operator and local governments. Such action is worth taking though, as the example of Wieliczka shows the effective competition among individual transport and railway is possible.

References

1. Hine, J.: Integration integration integration... Planning for sustainable and integrated transport systems in the new millennium. Transp. Policy 7(3), 173–175 (2000). Pergamon
2. Hull, A.: Integrated transport planning in the UK: from concept to reality. J. Transp. Geogr. 13(4), 318–328 (2005)
3. Janic, M., Reggiani, A.: Integrated transport systems in European union: an overview of some recent developments. Transp. Rev. 21(4), 469–497 (2001)
4. NEA, OGM, University Of Oxford, Erasmus University, TIS.PT and ISIS: Integration and regulatory structures in public transport: Costs and Benefits of Public Transport Integration, Consulting report to DG-TREN, Interim Report I, NEA, Rijswijk (2003)
5. Preston, J.: What's so funny about peace, love and transport integration? Res. Transp. Econ. 29, 329–338 (2010)
6. Solecka, K., Żak, J.: Integration of the urban public transportation system with the application of traffic simulation. Transp. Res. Procedia 3, 259–268 (2014)
7. Communication from the Commission – The future development of the common transport policy. A global to the construction of a Community framework for sustainable mobility. COM (92) 494
8. Communication from the Commission to the Council and the European Parliament: Keep Europe moving – Sustainable mobility for our continent Mid – term review of the European Commission's 2001 Transport White Paper COM (2006) 314
9. Communication from the Commission: A sustainable future for transport: Towards an integrated, technology-led and user friendly system, COM (2009) 279
10. Resolution on the Commission Green Paper on The Citizens Network: fulfilling the potential of public passenger transport in Europe. COM (95)0601 – C4-0598/95
11. White paper European transport policy for 2010: time to decide. COM (2001) 370
12. White Paper – Roadmap to a Single European Transport Area – Towards a competitive and resource efficient transport system. COM (2011) 144 final
13. Capital Regions Integrating Collective Transport For Increased Energy Efficiency: Księga dobrych praktyk. Zbiór doświadczeń I poradnik dla decydentów. Projekt CAPRICE INTERREG IVC 2009 – 2011. http://www.caprice-project.info/IMG/pdf/GPG_polish_version.pdf
14. Ibrahim, M.F.: Improvements and integration of a public transport system: the case of Singapore. Cities 20(3), 205–216 (2003)
15. Preston, J., Marshall A., Tochtermann, L.: On the Move: delivering integrated transport in Britain's cities, Center for Cities, London (2008)

16. Pucher, J., Buehler, R.: Integrating bicycling and public transport in North America. J. Public Transp. **12**(3), 79–104 (2009)
17. Solecka, K.: Tools for urban transportation integration, Monografia 324: Contemporary transportation systems. Selected Theoretical and Practical Problems. New Mobility Culture. Wydawnictwo Politechniki Śląskiej, Gliwice, pp. 163–170 (2011)
18. Banister, D., Givoni, M.: Integrated Transport. from Policy to Practice. Routledge, London (2010)
19. Eggenberger, M., Partidário, M.R.: Development of a framework to assist the integration of environmental, social and economic issues in spatial planning. Impact Assess. Proj. Apprais. **18**(3), 201–207 (2000)
20. Preston J.: Integration for Seamless Transport, Discussion Paper. International Transport Forum (2012)
21. Integracja transportu pasażerskiego w Unii Europejskiej pod redakcją Bąk M., Zeszyty Naukowe Uniwersytetu Gdańskiego Ekonomika Transportu i Logistyka, vol. 45, Gdańsk (2012)
22. Ciastoń-Ciulkin, A., Puławska-Obiedowska, S.: Solutions for agglomeration railway integration – case study of the line Wieliczka – Krakow Airport. In: Macioszek, E., Sierpiński, G. (eds.) Contemporary Challenges of Transport Systems and Traffic Engineering. Lecture Notes in Networks and Systems, vol. 2. Springer, New York (2017)
23. Ciastoń-Ciulkin, A., Puławska-Obiedowska, S.: Konkurencyjność podsystemu kolei aglomeracyjnej na przykładzie połączenia Kraków – Wieliczka Zeszyty Naukowo-Techniczne Stowarzyszenia Inżynierów i Techników Komunikacji w Krakowie. Seria: Materiały Konferencyjne, pp. 7–22 (2016)
24. Matulski, D.: Analiza wielkości potoków pasażerskich korzystających z integracji autobusowych linii dojazdowych B1, D1 i R1 i połączenia kolejowego Wieliczka Rynek Kopalnia – Kraków Lotnisko, praca inżynierska pod kierunkiem Ciastoń-Ciulkin A., Politechnika Krakowska (2017)
25. Pomiar napełnienia w pociągach Kolei Małopolskich na trasie Krakow Lotnisko/Airport - Wieliczka Rynek-Kopalnia (Research on volume of passengers flows on the route Krakow Lotnisko/Airport - Wieliczka Rynek-Kopalnia) Final report, Koło Naukowe Logistyki TiLOG Politechnika Krakowska, Krakow (2016)
26. Road Infrastructure: Blue Book, New Edition. Joint Assistance to Support Projects in European Regions (2015)

Trends in Free Fare Transport on the Urban Transport Service Market in Poland

Katarzyna Hebel[(✉)], Marcin Wołek, and Aleksander Jagiełło

Chair of Transportation Market, Faculty of Economics,
University of Gdańsk, Gdańsk, Poland
k.hebel@zkmgdynia.pl, mwol@wp.pl, a.jagiello@ug.edu.pl

Abstract. In recent years, free urban transport has become the area of interest among numerous groups. The article presents the analysis of social, economic and environmental rationale for introducing free urban transport.

Moreover, the Polish national and local laws and regulations were analysed to assess the introduction of free urban transport for various social groups. Using the source information from the Urban Transport Chamber of Commerce the criterion of age was analysed as grounds for free urban transport in Poland.

Using the budget-related data from the Polish cities and municipalities from 2000 to 2016 the economic implications resulting from entitlements to free urban transport were identified.

The conducted analyses revealed current trends regarding the entitlements to free urban transport and indicated that the main reason involves introducing such entitlements for children and youth, in small and medium towns and large cities and for seniors of over 70 years of age.

Keywords: Urban transport · Free public transport · Transportation market

1 Introduction

Urban transport has been one of the most important sub-systems which determine the city attractiveness. Together with non-motorised transportation it constitutes grounds for the concept of sustainable urban mobility. Its competitiveness relative to individual transportation has decreased in the Polish cities. It is facilitated on the one hand, by liberal car-related transport policy and mode of planning the urban space, and on the other, regarding individual decisions on choosing particular mode of transport – by shorter travel time and comfort of travel [1].

Increased passengers' expectations towards quality of public transport creates strong pressure on public finances. One possible strategy is to improve cost coverage ratio and reduce subsidies [2].

But in many cases urban transport performs also the social function reflected by a wide range of reduced fares and right to free travel for selected social groups. The most radical solution in this respect is the so-called Free Fare Public Transport.

© Springer Nature Switzerland AG 2019
G. Sierpiński (Ed.): TSTP 2018, AISC 844, pp. 208–217, 2019.
https://doi.org/10.1007/978-3-319-99477-2_19

2 Rationale for Free Fare Public Transport

In recent years, free fare transport on the market of urban transport services constitutes the area of interest for many social and political circles in Poland. The most fundamental argument of the supporters of wide-scope free-fare urban transport is the possibility to increase the attractiveness of public transport relative to private cars. The social argument in favour of the introduction of free fare public transport indicates the possibility to reduce the so-called mobile exclusion of people with limited access to private cars as well as low income [3]. The social grounds for the introduction of free fare urban transport may also involve family-friendly policy. The improvement of economic situation of households may result from the fact that the city travel expenses constitute an important share in their budgets and they increase with the number of children.

At the same time, it is estimated that the economic impact resulting from the implementation of free fare urban transport can be acceptable since the share of farebox revenues in financing the collective public transport in the Polish cities amounts between 20–40%. Moreover, it would be possible to reduce the costs of ticket distribution, which amounts to several percent in the budgets of transport operators. Furthermore, it is indicated that the costs of collective transport in the cities are relatively low taking account of municipal budgets, compared to the costs of other operations (e.g. support of educational system, municipal services) [3].

The economic rationale for free fare urban transport includes competition with neighbouring municipalities for tax payers. When free fare transport is introduced for the inhabitants of a particular municipality, the inhabitants of neighbouring municipalities may change their place of residence (at least formally) to obtain the right to free travel, which at the same time usually results in the need to pay taxes in the municipality providing free fare public transport for their inhabitants (e.g. Tallin in Estonia).

Another rationale for free fare public transport involves environmental aspect, namely efforts to reduce exhaust emissions through limiting the number of car travels in the city. Free fare public transport is regarded as a travel demand measure among traffic tolls, car-free streets and increased fuel taxes [4] and reducing congestion.

Free fare urban transport is often introduced for limited period of time (e.g. completion of particular investment), trial period, as part of research process or in emergency situations (flood in Prague in 2002, big events).

Another rationale involves promotion and education related to the desirable travel behaviour. For this purpose, some groups of people, e.g. holders of driving license obtain the right to free fare travel by public transport on a particular day (such campaign frequently coincides with the European Week – Sustainable Transport).

Consequently, there are a number of grounds for the decision to introduce free fare urban transport. Regardless of the aforementioned reasons, the decision to introduce the analysed solution always has a political dimension which becomes extremely important during the election campaign.

3 Overview of Practical Experience in Free Fare Public Transport (FFPT)

The first European city which, in 1997, introduced free fare urban transport for all the inhabitants is the Belgian city of Hasselt with ca. 76 thousand of inhabitants. As a result, the demand for urban transport services increased fourfold during the first year. In subsequent years, transport services increased and stabilized in 2005 [5]. Such significant increase in the demand saw the necessity to increase the supply, which in turn increased the Hasselt city budget expenditure to maintain the collective transport threefold during that time. Therefore, the absolute free fare public transport was abandoned towards the introduction of 60 eurocent fare and free fare urban transport only for the young under 19 years of age [6].

The studies on the so-called additional passengers in Hasselt showed that 16% of people gave up private car for public transport, the rest were people who replaced walking (9%) and cycling (12%) with the collective public transport. The other part of increased demand resulted from the increased mobility of people who used collective transport before, which is not beneficial from the perspective of sustainable urban mobility philosophy [7] formed by the so-called "non-motorised" forms of transportation (walking and cycling).

The second city which, in 2013, decided to introduce free fare collective transport is the Estonian city of Tallin. Since the collective transport service fares, before introducing free fare travel, were rather low and the share of collective transport services relatively high, the city observed only 5% increase in the share of collective public transport services, in the so called modal split, relative to private car transport, by 3 percentage points, whereas pedestrian traffic decreased from 12 to 7% [8]. Already when the decision was taken it was indicated that there was a threat for the electric transport subsystem operations in the long run due to the lower level of income (Tallin has tram and trolleybus transport).

The literature review shows a danger of a large spillover of benefits to particular groups which is far from fundamental social justice principle [9].

In Poland, in 2014 the absolute free fare urban transport was introduced in Żory (ca. 60 thousand inhabitants). Similar solutions were implemented in other cities, although they differ in the scope of and rationale for their implementation [10].

4 Methodology

The article provides the descriptive analysis of the national laws and regulations related to the entitlement to free fare urban transport. Then, the municipal (local) laws and regulations were evaluated under comparative analysis regarding the same theme. The source material comprised information from the Urban Transport Chamber of Commerce in Warsaw and information from particular municipalities.

In Poland, there are two types of entitlements to free fare public transport. The first results from statutory rights which shall be applied in all urban municipalities (cities) and municipal entitlements which can be established independently by local authorities, depending on the needs and will of those municipalities and first of all their financial standing.

4.1   Scope of Statutory Entitlements to Free Fare Public Transport Services in Poland

The scope of statutory entitlements to free fare public transport services in Poland is not very wide. The scope covers:

- MPs and senators [11],
- war and military invalids and accompanying carers of war and military invalids included in the 1st class of invalidity [12].

Moreover, municipalities shall provide free fare transport and care or refund of transport costs of a child and child carer by public transport [13].

- if transportation is provided by parents, when the distance from home to pre-school facility exceeds 3 km:
  - to children aged 3–7 under pre-school education;
  - to children aged 7–9 holding certificate of special education under pre-school education,
- when the distance to school exceeds 3 km: to primary school pupils of 4th grade,
- when the distance from home to school exceeds 4 km – to primary class pupils of 5th to 7th grade and middle school pupils.

The above mentioned complex regulations involve the responsibility to locate schools so that the distance to the nearest school fails to exceed the specified 3 km for the youngest children and 4 km for older children. If the conditions in a particular municipality are satisfied – there is no need to provide free fare public transport to schools. Therefore, in large cities with usually numerous schools, the above mentioned regulation is not used. In means the necessity to organize the transport for children in the majority of rural municipalities of low population density.

In smaller municipalities, agreements are concluded with various entities, e.g. taxi corporations which transport selected children to schools and submit invoices for their services directly to municipalities.

However, instead of calculating the distance accurately, very often pupils are granted entitlements to free fare public transport, not only on their way to school but also for all travels.

We need to add that statutory regulations in Poland also introduce entitlements to reduced fare transport (50% discount compared to standard price) for:

- war veterans and victims of oppression [14],
- students (people at bachelor studies and master studies or uniform master studies) [15],
- young people included in the 1st class of invalids who are not older than 30 years of age [16],
- blind victims of military actions and their carers [17].

4.2 Scope of Municipality (Local) Entitlements to Free Fare Public Transport Services in Poland

The other group of entitlements to free fare urban transport includes municipal entitlements. The analysis conducted under the data from municipalities of various size (members of the Urban Transport Chamber of Commerce in Warsaw) makes it possible to differentiate the typical groups of the entitled, including:

- children up to 4 years of age;
- people who reached a certain age, usually 70 years;
- pensioners and disability pensioners;
- severely disabled persons and their accompanying carer (adult);
- severely or moderately visually-impaired people and their accompanying carer;
- disabled children up to 7 years of age and their accompanying carer;
- disabled children attending kindergarten or pre-school facilities in primary schools;
- disabled pupils;
- Volunteer Blood Donors of the Polish Red Cross;
- persons employed at the urban transport operators and carriers in particular municipalities;
- members of large families.

Apart from the typical groups of people entitled to free fare urban transport, we can indicate less popular and even atypical groups. The related example solutions were presented in Table 1. The atypical groups include students since, as previously mentioned, they are entitled to reduced fare urban transport in all Polish municipalities under statutory regulations.

Table 1. Groups entitled to free fare collective urban transport in the Polish municipalities

People entitled to free fare collective urban transport	Example municipality applying particular solution
Everyone	Żory
Municipality inhabitants	Mława
People born in a particular municipality	Augustów
Honorary inhabitants of a particular municipality	Tarnów
Tourists (tourist card holders)	Augustów
The unemployed	Rzeszów
Newly employed people (first three months after finding employment)	Gdańsk
Drivers (under vehicle registration document and motor vehicle liability insurance)	Zakopane
Primary school pupils	Częstochowa
Post-primary school pupils	Kołobrzeg
Students	Jastrzębie-Zdrój
Large family card holders	Gdynia
Children under social care	Tychy

Source: Own elaboration

The thorough analysis has been conducted relative to the age of seniority as a criterion which entitles to free fare public transport. The analysis covered 100 Polish cities of the largest number of inhabitants, i.e. from 74.1 thousand (Piła) to 1.75 million (Warsaw). The analysis proves that:

- in 4 of them (Lubin, Żory, Bełchatów, Legionowo) public transport is free for all passengers.
- other cities offer entitlements to free fare travel for passengers over a particular age; in 4 cities the age limit which entitles to free fare travel is 65 years and in one (Stalowa Wola) – even 60 years (Fig. 1).

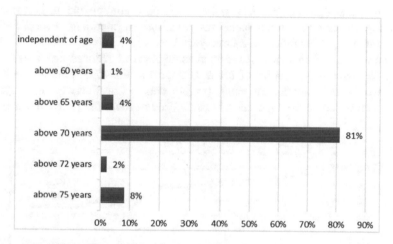

Fig. 1. Age criterion which entitles seniors to free fare public transport in 100 largest Polish cities. Source: Own elaboration

The seniority as a criterion which entitles to free fare public transport shall be analysed together with seniors' social status. In Poland in October 2017, regulations on eligibility for the retirement age pension were amended. The previous age of 67 for both, men and women, was replaced with 60 years for women and 65 years for men. It means more people who retire and become entitled to reduced fare urban transport in the majority of Polish municipalities than before. It may hamper the tendency to reduce the age which entitles to free fare urban transport.

The scope of entitlements to free fare urban transport established by particular municipalities is not uniform. We can differentiate the following scopes:

- spatial scope (area of administrative unit or its part);
- functional scope (particular line/urban transport lines - In Kielce free fare OW and OZ lines were introduced to revive traffic in the city centre. In Wodzisław zero rate is applicable to the line connecting the bus station with the railway station);
- time-specific scope (days in week, particular day);
- subjective scope (people meeting particular criteria).

Statutory regulations as well as municipal regulations provide entitlements to free travel; they mainly refer to students and pensioners, but there is little awareness among the Polish society that the entitlements to free fare urban transport for students and pensioners result from local and not national regulations.

5 Economic and Social Implications of Entitlements to Free Fare Urban Transport in Poland

The current budgetary expenditure of the Polish local government authorities is very restrained. The municipal budget provides financing for entities performing such tasks as education and upbringing, social policy, transport, administration, and municipal engineering. In practice, current spendings constitute an important element for calculating the municipal liabilities on a three-year basis.

In the area of urban transport the increasing level of reduced and free fare with pressure to increase the quality of urban transport at the same time, current demographic trends and patterns for using private cars – constitute grounds for rapid decrease in the income from the sale of tickets used to cover the expenditures (Fig. 2).

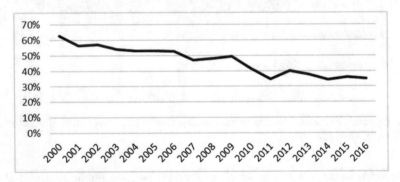

Fig. 2. Farebox revenues to costs of public transport ratio [%] (cost recovery) in selected Polish cities between 2000 and 2016. Source: Own elaboration based on the data of the Urban Transport Chamber of Commerce

The introduction of free fare urban transport should always be preceded by cost analysis. One of the methods involves assessing the value of lost income from the sale of tickets and treating the value as an equivalent for the costs of introducing free fare urban transport.

Other possible economic implications of free fare urban transport include:

- risk of surge in the level of demand for urban transport services;
- loss of possibility to shape urban transport service supply under market economy principles;
- increase in the susceptibility of urban transport system to budgetary disturbances (from 1991 Poland has not experienced crisis of structural character);

- shaping claimant attitude among other groups of people, not included in the previous system of free fare urban transport;
- risk of abandoning other instruments for spatial and transport order development.

Providing particular social and professional groups with entitlement to free fare urban transport we need to take account of very significant trend related to ageing of the Polish society and the fact that apart from employed persons the largest group using collective transport services is youth at school (Fig. 3).

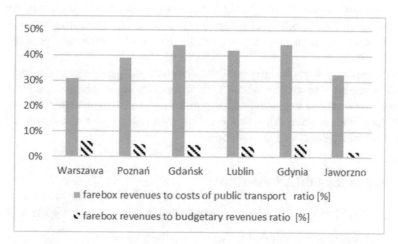

Fig. 3. Cost coverage ratio and farebox revenues to total budgetary revenues in selected Polish cities in 2016

By means of price policy in urban transport not only economic but also political objectives are achieved. As a result, the scope of entitlements to free fare urban transport changes along with the objectives the municipality intends to achieve. The sign indicating attempts to support the unemployed and encourage them to look for job actively is to grant them right to free fare urban transport, which seems pointless in the Internet age with job offers posted regularly online.

Another example of social policy refers to entitlements for children and youth as well as members of large families.

6   Trends in Granting Entitlements to Free Fare Urban Transport in the Polish Cities

The trends in granting entitlements to free fare urban transport in the Polish cities result from several crucial factors. The main factor involves the evolution of objectives of widely understood social policy.

Changes in providing free fare urban transport occur because the objectives behind these entitlements change.

By following the objectives of family friendly policy, the authorities of more and more cities grant such entitlements to children aged 4 or 7 and students, including mainly primary school pupils but also students of middle and post-middle schools. Another instrument used to achieve the family friendly policy objectives involves granting the entitlement to free fare urban transport to all members of large families, not only children.

Yet another observed trend indicating changes in granting entitlement to free fare urban transport involves lowering the age of the entitled. Granting such entitlements alone results from actions striving to fight social exclusion of senior citizens for whom urban transport fares with decreasing income resulting from the termination of employment are becoming a very significant financial burden and psychophysical difficulty.

The transport policy objectives can be achieved via granting entitlements to free fare urban transport to drivers who - with the possibility of choosing their mode of transport – own private car or collective public transport – choose public transport. It aims at reducing the congestion in the cities, although so far we lack research confirming the effectiveness of such solution.

7   Conclusions and Discussion

The article has identified the following trends in free fare urban transport in Poland:

- The number of people entitled to free fare public transport has been increasing;
- The entitlements to free fare travel most frequently cover children, not only pre-school children, but also pupils of primary and post-primary schools;
- The second largest group of people entitled to free fare urban transport are seniors, with the senior's age limit usually amounting to 70 years, although there are cities where seniors are granted such entitlement earlier;
- As a result of facilitated use, most frequently the entitlements refer to the entire city area and all days in a week throughout the whole day;
- Cities with free fare public transport for all inhabitants are still rare across the country.

The rationale for introducing free fare urban transport include mainly social and political reasons, in particular family friendly policy and pro-environmental policy (in cities with smog issues). The rationale resulting from transport policy is less important.

Apart from the rationale for introducing free fare urban transport presented in this article it would be worth analysing such grounds as the promotion of collective transport [18], whereas the economic analysis of the implications of such entitlements could be extended by the evaluation of the impact of complementary service prices for people using private cars.

References

1. Hebel, K., Wołek, M., Wyszomirski, O.: Rola samochodu osobowego w podróżach miejskich mieszkańców Gdyni w świetle wyników badan marketingowych w 2015 roku. Przegląd Komunikacyjny **2**, 21–24 (2017)
2. van Goeverden, C., Rietveld, P., Koelemeijer, J., Peeters, P.: Subsidies in public transport. Eur. Transp. **32**, 6 (2006)
3. Tomanek, R.: Rola bezpłatnego transport zbiorowego w równoważeniu mobilności w miastach. Transp. Miejski i Regionalny **7**, 5 (2017)
4. Robèrt, M., Jonsson, R.D.: Assessment of transport policies toward future emission targets: a backcasting approach for Stockholm 2030. J. Environ. Assess. Policy Manag. **8**(4), 452 (2006)
5. Boussauw, K., Vanotrive, T.: Transport policy in Belgium: translating sustainability discourses into unsustainable outcomes. Transp. Policy **53**, 11–19 (2017)
6. Canters, R.: Hasselt cancels free public transport after 16 years (Belgium). Eltis (2014) http://eltis.org/discover/news/hasselt-cancels-free-public-transport-after-16-years-belgium-0. Accessed 30 Mar 2018
7. Fearnley, N.: Free fares policies: impact on public transport mode share and other transport policy goals. Int. J. Transp. **1**(1), 82 (2013)
8. Cats, O., Susilo, Y.O., Remail, T.: The prospects of fare-free public transport: evidence from Tallin. Transportation **44**, 1083–1104 (2016). https://doi.org/10.1007/s11116-016-9695-5
9. Philipson, M., Willis, D.: Free public transport for all? In: Paper Presented on the Australian Transport Forum, p. 633 (1990). http://atrf.info/papers/1990/1990_Philipson_Willis.pdf
10. http://www.transport-publiczny.pl/watki/bezplatna-komunikacja.html. Accessed 30 Mar 2018
11. Act of 9 May 1996, on the performance of the deputy and senator's mandate. Journal of Laws 2016 r., item 1510 with further amendments
12. Act of 29 May 1974, on supplying war invalids and military invalids and their families, Journal of Laws 2017, item 2193
13. Act of 7 June 1991, on the education system, TJ. Journal of Laws 2017, item 2198 with further amendments
14. Act of 24 January 1991, on the combatant and certain victims of repressions in wartime and in the post-war period. Journal of Laws 2018, item. 276
15. Act of 27 July 2005, law on higher education. Journal of Laws No 2017, item 2183 with further amendments
16. Regulation of the Council of Ministers of 1st of June 1987 on determining the rules and procedure for keeping records, as well as the scope and forms of assistance for young people affected by invalids and young people with disabilities, Dz. U. 1987, Nr 23, poz. 130
17. Act of 16 November 2006 r. on providing help and entitlements for civil, blind war victims. Journal of Laws No 2006., No 249, item 1824
18. Bąkowski, W.: Ekonomiczne podejście do problemu taryfy zerowej w komunikacji miejskiej. Komunikacja Publiczna **1**(70), 25–29 (2018)

Urban Transport Integration Using Automated Garages in Park and Ride and Car-Sharing Systems – Preliminary Study for the Upper Silesian Conurbation

Grzegorz Sierpiński[(⊠)], Katarzyna Turoń, and Czesław Pypno

Faculty of Transport, Silesian University of Technology, Katowice, Poland
{grzegorz.sierpinski,katarzyna.turon,
czeslaw.pypno}@polsl.pl

Abstract. The transport system in many cities is becoming overloaded due to the increasing number of travels, a large share of which involve passenger cars. One of the solutions improving transport system functioning is based on a concept of changing the behaviour of travellers and of increasing the share of travels using means of transportation dedicated by the city (urban public transport, city bicycle and car rental systems). This change, even if implemented only on some travel sections, may lead to a substantial reduction of the negative impact of transport on the environment (in terms of emissions as well as of the space occupied). The paper proposes to integrate transport in cities by using automated multi-storey large-capacity garages in certain locations. A case study is also presented here for the Upper Silesian Conurbation, describing the potential impact on the transport system in that urban area. The proposed solution may also be used in different areas.

Keywords: Automated garage · Park and ride · Car-sharing

1 Introduction

The compactly built-up urban setting restricts the possibilities of expanding the transport network in cities. Given the constant increase of the number of travels in cities, solutions continue to be sought aimed at increasing the capacity of the transport network in terms of the number of passengers carried rather than of the number of vehicles, with the simultaneous reduction of the negative impact of transport on the surrounding environment [1–4]. One way of improving the situation consists in an attempt to change travellers' behaviour by increasing the share of travel using public transport or city bike and car rental systems, or even that of travelling on foot (e.g. [5, 6]). The choice of the travel method depends on many factors, while a clear message with regard to transport organisation translates into the appeal of the given area using the specific means of transport (periodic research is carried out with regard to this, making it possible to identify the factors having the greatest impact on travellers' behaviour, e.g. [7–13]). It is often difficult to achieve a complete modal shift from passenger cars to public transport. Consequently, efforts are made to integrate and

G. Sierpiński (Ed.): TSTP 2018, AISC 844, pp. 218–228, 2019.
https://doi.org/10.1007/978-3-319-99477-2_20

combine travels into chains, with the consecutive sections covered in different ways [14, 15]. This trend has been taken from the supply chain policy related to freight transport [16–18]. In this case, travel usually starts from one's place of residence using a private passenger car. This is followed by switching to a more environmentally friendly means of transportation offered by the city, at a specific location. Car parks need to be built at the modal shift points, and this requires a sufficiently large amount of space. In order to make the modal shift possible, urban transport must be available in close proximity to the car park. In such cases, a park-and-ride system is most often set up, in the form of an integrated hub [19]. The car park can have any form, but it is the main factor that determines spatial occupancy as well as the appeal of the given solution. The most advantageous solution today is a parking system that can be integrated into the city's Intelligent Transport System (ITS). This form contributes to better traffic organisation by way of the additional clear information provided to travellers (using variable-message signs and on-board information systems). It is also worth mentioning that a city with properly organised parking spaces is attractive for real estate developers, tourists, and also the local inhabitants [5, 20, 21].

The paper presents a concept of integrating urban transport using automated garages. The system is based on a patented technical solution involving large-capacity garages. A preliminary study was conducted for the area of 19 cities belonging to the Upper Silesian Conurbation (Poland). The traditional approach that involves placing park-and-ride systems at city borders due to the dense transport network has been replaced here by more user-friendly locations at public transport stops with a large number of connections to other places within the area.

2 Automated Multi-storey Garages

Over the recent years, an increasing number of buildings with underground multi-storey garages have been built in city centres. They have no more than four or five storeys, with a half-storey ramp structure. In the fully-automated above-ground garages built today, mechanical parking systems make it possible to park or to pick up the car within 40–120 s, depending on the storey, which corresponds to 30–90 cars per hour. On average, 60 cars per hour are assumed [22].

The proposed concept of integrating individual and collective transport proposes an above-ground multi-storey automated garage with a high capacity of 400 cars per hour in both directions [23, 24]. In the basic version, the garage built on a rectangular plan has 17 storeys with parking spaces, with the ground floor used as a loading/unloading area. There are 60 parking spaces on each storey. The column-and-beam supporting structure of the garage is made of rolled and bent steel profiles. The whole structure rests on a reinforced concrete voided foundation slab, in which some of the mechanical equipment can be installed. For instance, the ground floor machinery includes scraper-plate conveyors which receive and return the cars. From the loading zone at level 0, the cars are transported upwards on pallets to the individual storeys by the first electric lift (Fig. 1). On each storey, mobile platforms take the cars from the lifts and carry them to the individual parking stalls. The cars are transported back down to level 0 by means of another electric lift.

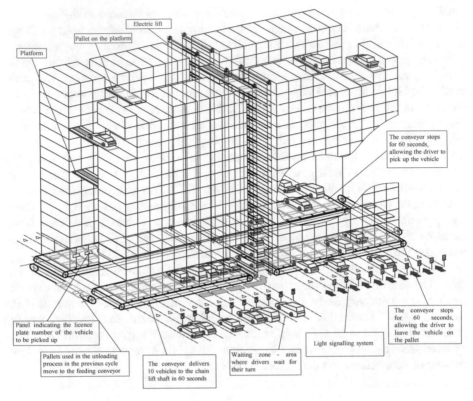

Electric lift

Pallet on the platform

Platform

The conveyor stops for 60 seconds, allowing the driver to pick up the vehicle

The conveyor stops for 60 seconds, allowing the driver to leave the vehicle on the pallet

Light signalling system

Panel indicating the licence plate number of the vehicle to be picked up

Pallets used in the unloading process in the previous cycle move to the feeding conveyor

The conveyor delivers 10 vehicles to the chain lift shaft in 60 seconds

Waiting zone - area where drivers wait for their turn

Fig. 1. General layout of the proposed automated garage (source: own research)

In this layout, the garage has 2×10 car entrances on the ground floor, allowing drivers to drive alternately onto the receiving conveyors and leave their cars. The whole process is regulated by a light signalling system. The maximum and safe time of entry for 10 cars onto the receiving conveyor is 90 s. During this time, a video detector reads the vehicle registration number and sends it to the central garage control system, and the driver leaves the vehicle. After that, within the next 90 s of garage operation, the lift transports the cars upwards (in nine-second cycles), and then the cars are gradually taken to the stalls on the individual storeys. At the same time, during the same 90 s of garage operation, new vehicles drive onto the second receiving conveyor. In this way, the garage is operated on an alternating basis, and vehicles are distributed among the storeys as other vehicles enter and exit the garage [25].

The operation of the conveyors receiving and returning the cars is synchronised with the operation of the lifts transporting the cars upwards and downwards, as well as with the operation of the platforms. Once the conveyor receiving the cars for about 90 s is full, they start to be cyclically transferred to the lift, from which they are also cyclically moved on the platforms of each garage storey.

Although the example features a 17-storey garage, the ultimate structure may be flexibly adapted to the spatial resources of the given area. A significant advantage of this structure is the possibility of several cars entering and exiting at the same time, which significantly reduces the waiting time and substantially improves capacity.

3   Case Study in the Upper Silesian Conurbation Area

3.1   Area Description

The Upper Silesian Conurbation area was chosen for the preliminary case study. The area includes 19 cities. It has a very dense transport network and it is one of the most densely populated areas in Poland. According to the research and analyses carried out by the authors within the framework of the Green Travelling project, more than 4.7 million travels take place every day in the area, with a total population of more than 2.2 million. This number requires resolute action in terms of urban transport organisation and traffic stream management.

Several carriers provide public urban transport services in the studied area, including the following:

- KZKGOP, providing services using bus and tram lines (total number of stops in the analysed area: 1992 [26]);
- PKM Jaworzno, providing bus transport services (total number of stops in the analysed area: 238 [27]);
- MZK Tychy, with bus and trolleybus lines (total number of stops in the analysed area: 381 [28]).

Bike-sharing and car-sharing systems are also available in the area.

The bike-sharing operator in the Upper Silesian Conurbation is Nextbike, a German company. Its activities carried out in selected cities in the conurbation area include the provision of dedicated bike-sharing services, with the following systems [29]:

- City by bike – Katowice, operating since 1 May 2015, offering 322 bikes and 40 docking stations;
- GRM – Gliwice City Bicycle – Gliwice, operating since 28 March 2017, offering 100 bikes and 10 docking stations;
- Nextbike Sosnowiec – Sosnowiec, operating since 30 March 2018, offering 130 bikes and 9 docking stations.

The bike-sharing systems available in the conurbation are very popular among users. In 2017, the bicycles available from the City by Bike system in Katowice were hired 103,620 times [30]. For the sake of comparison, they were hired 38,217 times in 2016 [31].

Since 2017, it has also been possible to use car-sharing services offered by GreenGoo in the conurbation area. The system offers only a few vehicles, but users can choose from the following fully electric cars: Nissan Leaf, Renault Zoe, BMW i3, and Tesla Model S60 [32]. On top of that, in March 2018, the pioneer of car-sharing in Poland, Traficar, entered the Upper Silesia and Dąbrowa Basin market. The system operates in the following cities in the conurbation: Będzin, Bytom, Chorzów, Dąbrowa Górnicza, Gliwice, Jaworzno, Katowice, Mikołów, Mysłowice, Ruda Śląska,

Siemianowice Śląskie, Sosnowiec, Tychy, and Zabrze [33]. One can also pick up and return a car at Katowice Airport in Pyrzowice [33]. Currently, 300 Renault Clio vehicles are available to users [33]. They can be parked in any public parking space in the cities covered by the system. Moreover, parking spaces dedicated to car-sharing are gradually appearing in the cities, which may prove that more parking space, including automated garages, needs to be provided. The availability of special parking spaces dedicated to car-sharing in Katowice is presented in the Traficar application in Fig. 2.

Fig. 2. Availability of parking spaces dedicated to car-sharing in the area of the city of Katowice (source: Traficar app.)

3.2 Identification of Areas to Implement the Solution

As part of the studies, it was proposed that a modal shift should be allowed near bus stops characterised by the largest number of departures of public transport vehicles on weekdays and Sundays. One such bus stop was identified for this purpose in each of the 19 cities in the studied area. The location of the bus stops is shown in Table 1 and in Fig. 3.

The places where the vehicles could be left should be located not far from public transport stops or city bike rental stations. It is not easy to find an optimum location for the car park, as there is a large number of factors that can influence its attractiveness (e.g. [34, 35]). The literature indicates the maximum access distances, ranging from 300 to 960 [m]. This is quite a significant difference. Long distances were proposed in [36], where the distance to the bus stop was identified at 640 [m], and for tram stops, it was 960 [m]. Shorter distances were usually assumed in other studies, often 300 [m] for bus stops and 400 [m] for tram stops (for example [37]). In order to determine travellers' preferences in the analysed area, transport studies were carried out in relation to

Table 1. List of selected stops in the area of 19 cities.

No. of city	City name	PuT stop name	Coordinates
1	Katowice	Katowice Dworzec	50.25849, 19.01763
2	Chorzów	Chorzów Estakada	50.29655, 18.95517
3	Zabrze	Zabrze Goethego	50.30562, 18.78276
4	Gliwice	Gliwice Plac Piastów	50.29828, 18.67649
5	Mysłowice	Mysłowice Dworzec PKP	50.23844, 19.14252
6	Sosnowiec	Sosnowiec Dworzec PKP	50.27808, 19.12699
7	Dąbrowa Górnicza	Dąbrowa Górnicza Centrum	50.32529, 19.1879
8	Czeladź	Czeladź Stare Miasto	50.31925, 19.07043
9	Będzin	Będzin Dworzec PKP	50.31844, 19.13341
10	Bytom	Bytom Dworzec	50.34388, 18.91338
11	Ruda Śląska	Kochłowice Kościół	50.25407, 18.90085
12	Świętochłowice	Świętochłowice Mijanka	50.28621, 18.92863
13	Siemianowice Śląskie	Bytków Osiedle Chemik	50.2973, 19.00394
14	Tychy	Tychy E.LECLERC	50.12748, 18.97394
15	Mikołów	Mikołów Dworzec PKP	50.17206, 18.90052
16	Knurów	Knurów Szpitalna	50.22707, 18.65743
17	Tarnowskie Góry	Tarnowskie Góry Dworzec	50.44809, 18.86423
18	Piekary Śląskie	Osiedle Wieczorka Dworzec	50.39174, 18.96071
19	Jaworzno	Jaworzno Centrum	50.20338, 19.27121

Source: Own research

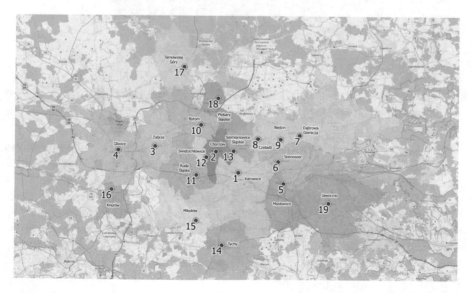

Fig. 3. Map of the Upper Silesian Conurbation area with the 19 cities and the identified garage locations marked (source: own research)

them. The research was conducted within the framework of the Green Travelling project, as part of the ERA-NET Transport III: Future Travelling programme [38]. The mean identification value obtained was 415 [m], while for the tram stop the distance was longer, i.e. 501 [m].

Travellers indicated the maximum access distance for the tram stop as one ranging from 400 to 600 [m] (24.6% of the respondents) and from 200 to 400 [m] for the bus stop (25.2% of the respondents). Figure 4 shows the location of the proposed points as well as the public transport routes. Given that the proposed automated garage is a flexible structure in terms of the total number of parking spaces, as it has already been mentioned, it is possible to match the size of the garage to the specific needs, as well as to the existing urban development.

Fig. 4. Map of the Upper Silesian Conurbation area with public transport routes and the identified garage locations marked (source: own research)

The idea of combining travels into chains in order to favour environmentally-friendly travel in general (and not within a specific city zone) makes it possible to look at the traveller's problem differently. There is no need to plan travelling with one's own vehicle to the outskirts of the city. In the proposed solution, the change is to take place only on a small section of the entire route. However, as a result of such changes, taking place in "small steps", travellers may develop a habit of seeking advantages in the modal shift much more rapidly [39]. At the same time, the availability of a car-sharing system in such places shortens the time of searching for them in the urban transport system.

3.3   Interpretation of Implementation and Potential Impact on the Transport System in the Area

Urban transport integration constitutes a major challenge for the local authorities. Due to the growing congestion problems and the need to reduce harmful emissions, cities have been implementing diverse infrastructural and organisational initiatives.

The solution proposed in this paper requires considerable financial outlays, which makes wider analyses necessary with regard to the siting of the proposed garages. The next step involves choosing the right incentive to draw the attention of travellers using private cars to this solution. The main incentive used in the case of park-and-ride systems is the possibility of parking one's car free of charge if a public transport ticket is purchased, or of obtaining a discount on public transport if a parking fee is paid (e.g. [40–42]). At the same time, drivers using such garages may receive a discount for entering the zone in their immediate vicinity (if there is a paid area) [43]. An incentive is also required for car-sharing system users, to make them more willing to choose the automated garage as the place where they leave their cars. A discount on rental may also have a positive effect here.

The development of Intelligent Transport Systems is possible with the full integration of various infrastructure elements and the vehicles [44]. Automated garages constitute a good source of information about the needs, and since they are centrally controlled, they enable a constant and dynamic flow of information about the current status of the garage.

Bearing the above in mind, the subsequent stages of the process of implementation of automated car parks as elements of the park-and-ride system have been identified (Fig. 5).

Fig. 5. Stages of implementation of the proposed solution (source: own research)

Possible benefits of the proposed solution include:

- reduction of harmful emissions (due to the fact that more environmentally-friendly means of transportation are used on a certain section when travelling);
- reduction of urban space occupancy (the vehicles in the garage are parked very close to one another and do not occupy additional on-street parking spaces);
- reduction of congestion in selected areas of the transport network (drivers find parking spaces more quickly, i.e. in the garage, and continue their travel using public transport, thus reducing the time when private vehicles occupy urban space).

It should also be taken into account that in the automated multi-storey garage discussed here, the car is received, loaded, unloaded and returned with the engine switched off. In other types of multi-storey car parks, drivers use a lot of fuel as they search for a free parking space, driving in low speed. It may be even more difficult to leave the car park during rush hour, when vehicles start to queue up in the circulation areas. Such situations generate additional harmful emissions, which is not the case in the proposed solution.

4   Conclusions and Further Research

Although the erection of enclosed garage buildings may give rise to controversies due to the costs, the idea should be consistently pursued in order to reduce on-street parking, which is also burdened with the risk of theft or damage to the vehicle, and often significantly disrupts pedestrian and cyclist streams.

In the concept presented in this paper, automated multi-storey garages have two main purposes – to ensure high availability of parking spaces for private passenger cars (which reduces the transport network occupancy on the part of such vehicles looking for a place to park), as well as high availability of city cars (which are usually more environmentally friendly). The holistic approach presented here makes the garage a more versatile facility, serving more than one group of users.

The next step related to the current trend in e-mobility development should involve promoting this solution also through the possibility of charging the vehicle while parked in the automated garage [45] and the integration of car-sharing systems with dedicated parking spaces in automated garages.

Acknowledgements. Selected of the present research has been financed from the means of the National Centre for Research and Development as a part of the international project within the scope of ERA-NET Transport III Future Travelling Programme "A platform to analyze and foster the use of Green Travelling options (GREEN_TRAVELLING)".

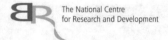

References

1. White Paper: Roadmap to a single European transport area – towards a competitive and resource efficient transport system. COM 144 (2011)
2. Communication from the Commission to the European Parliament, the Council, the European Economic and Social Committee and the Committee of the regions: clean power for transport: a European alternative fuels strategy. COM (2013) 17, Brussels, 24 January 2013
3. Banister, D.: The sustainable mobility paradigm. Transp. Policy **15**, 73–80 (2008)
4. Jacyna, M., Żak, J., Jacyna-Gołda, I., Merkisz, J., Merkisz-Guranowska, A., Pielucha, J.: Selected aspects of the model of proecological transport system. J. KONES Powertrain Transp. **20**, 193–202 (2013)
5. Turoń, K., Czech, P., Juzek, M.: The concept of walkable city as an alternative form of urban mobility. Sci. J. Silesian Univ. Technol. Ser. Transp. **95**, 223–230 (2017)

6. Nosal, K.: Travel demand management in the context of promoting bike trips, an overview of solutions implemented in Cracow. Transp. Probl. **10**(2), 23–34 (2015)
7. Staniek, M.: Moulding of travelling behaviour patterns entailing the condition of road infrastructure. In: Macioszek, E., Sierpiński, G. (eds.) Contemporary Challenges of Transport Systems and Traffic Engineering, Lecture Notes in Networks and Systems, pp. 181–191 (2017)
8. Macioszek, E., Lach, D.: Analysis of the results of general traffic measurements in the West Pomeranian Voivodeship from 2005 to 2015. Sci. J. Silesian Univ. Technol. Ser. Transp. **97**, 93–104 (2017)
9. Macioszek, E.: The comparison of models for follow-up headway at roundabouts. In: Macioszek, E., Sierpiński, G. (eds.) Recent Advances in Traffic Engineering for Transport Networks and Systems. LNNS, vol. 21, pp. 16–26. Springer, Cham (2018)
10. Celiński, I.: Transport network parameterisation using the GTAlg tool. In: Macioszek, E., Sierpiński, G. (eds.) Contemporary Challenges of Transport Systems and Traffic Engineering, Lecture Notes in Networks and Systems, pp. 111–123 (2017)
11. Piecha, J., Staniek, M.: The context-sensitive grammar for vehicle movement description. In: Bolc, L., Tadeusiewicz, R., Chmielewski, L.J., Wojciechowski, K. (eds.) Computer Vision and Graphics. Lecture Notes in Computer Science, vol. 6375, Part II, pp. 193–202. Springer, Berlin (2010)
12. Staniek, M.: Stereo vision method application to road inspection. Balt. J. Road Bridge Eng. **12**(1), 38–47 (2017)
13. Galińska, B.: Multiple criteria evaluation of global transportation systems-analysis of case study. In: Sierpiński, G. (ed.) Advanced Solutions of Transport Systems for Growing Mobility. AISC, vol. 631, pp. 155–171. Springer, Cham (2018)
14. Okraszewska, R., Romanowska, A., Wołek, M., Oskarbski, J., Birr, K., Jamroz, K.: Integration of a multilevel transport system model into sustainable urban mobility planning. Sustainability **10**(2), 479 (2018)
15. Stanley, J.: Land use/transport integration: starting at the right place. Res. Transp. Econ. **48**, 381–388 (2014)
16. Kijewska, K., Małecki, K., Iwan, S.: Analysis of data needs and having for the integrated urban freight transport management system. Commun. Comput. Inf. Sci. **640**, 135–148 (2016)
17. Macioszek, E.: First and last mile delivery-problems and issues. In: Sierpiński, G. (ed.) Advanced Solutions of Transport Systems for Growing Mobility. AISC, vol. 631, pp. 147–154. Springer, Cham (2018)
18. Wasiak, M., Jacyna, M., Lewczuk, K., Szczepański, E.: The method for evaluation of efficiency of the concept of centrally managed distribution in cities. Transport **32**(4), 348–357 (2017)
19. Song, Z., Hea, Y., Zhang, L.: Integrated planning of park-and-ride facilities and transit service. Transp. Res. Part C **74**, 182–195 (2017)
20. Saelens, B.E., Sallis, J.F., Frank, L.D.: Environmental correlates of walking and cycling: findings from the transportation, urban design, and planning literatures. Ann. Behav. Med. **25**(2), 80–91 (2003)
21. Montgomery, C.: Happy City: Transforming Our Lives Through Urban Design. Ferrar, Straus and Giroux, New York (2013)
22. Roodbergen, K.J., Vis, I.F.A.: A survey of literature on automated storage and retrieval systems. Eur. J. Oper. Res. **194**(2), 343–362 (2009)
23. Pypno, Cz.: Multi-storey automated over ground garage for cars - solution for parking problems in big urban areas. US Patent 4,039,089, WO 91/16515

24. Pypno, Cz.: Multi-storey, overground automated garage for parking cars and the way of taking and returning cars. Patent registration PL 216743
25. Pypno, C., Sierpiński, G.: Automated large capacity multi-story garage—concept and modeling of client service processes. Autom. Constr. **81C**, 422–433 (2017)
26. KZK GOP. http://www.kzkgop.com.pl/
27. PKM Jaworzno. http://www.pkm.jaworzno.pl/
28. MZK Tychy. http://www.mzk.pl/
29. NEXTBIKE. https://nextbike.pl/o-nextbike/
30. CITY BIKE. https://citybybike.pl/news/ponad-100-tysiecy-wypozyczen-city-by-bike/
31. CITY BIKE. https://citybybike.pl/news/oficjalne-podsumowanie-najlepszego-sezonu-w-historii-rowerow-miejskich-city-by-byke-170-dni-65-tysiaca-nowych-uzytkownikow-i-blisko-40-tysiecy-wypozyczen/
32. NOWA ENERGIA, GREENGOO. http://nowa-energia.com.pl/2017/09/25/punkt-ladowania-i-carsharingg-samochodow-elektrycych-otwarcie-w-galerii-katowickiej/
33. TRAFICAR. http://www.traficar.pl
34. Clayton, W., Ben-Elia, E., Parkhurst, G., Ricci, M.: Where to park? A behavioural comparison of bus park and ride and city centre car park usage in Bath, UK. J. Transp. Geogr. **36**, 124–133 (2014)
35. Holguin-Veras, J., Yushimito, W.F., Aros-Vera, F., Reilly, J.: User rationality and optimal park-and-ride location under potential demand maximization. Transp. Res. Part B **46**, 949–970 (2012)
36. Gent, C., Symonds, G.: Advances in Public Transport Accessibility Assessments for Development Control: A Proposed Methodology. Capita Symonds Ltd. Transport Consultancy, London (2005)
37. Loose, W.: Flächennutzungsplan 2010 Freiburg–Stellungnahme zu den verkehrlichen Auswirkungen. Öko-Institut eV, Freiburg (2001)
38. Green travelling: a platform to analyse and foster the use of green travelling options. Project Proposal, The ERA-NET Transport III: Future Travelling (2013)
39. Meyer, M.D.: Demand management as an element of transportation policy: using carrots and sticks to influence travel behavior. Transp. Res. Part A **33**, 575–599 (1999)
40. Nosal, K., Starowicz, W.: Evaluation of influence of mobility management instruments implemented in separated areas of the city on the changes in modal split. Arch. Transp. **35** (3), 41–52 (2015)
41. Buehler, R., Pucher, J.: Making public transport financially sustainable. Transp. Policy **18**, 126–138 (2011)
42. Paulley, N., Balcombe, R., Mackett, R., Titheridge, H., Preston, J., Wardman, M., Shires, J., White, P.: The demand for public transport: the effects of fares, quality of service, income and car ownership. Transp. Policy **13**, 295–306 (2006)
43. Szarata, A., Nosal, K., Franek, Ł., Duda-Wiertel, U.: The impact of the car restrictions implemented in the city centre on the public space quality. Transp. Res. Proc. **27**, 752–759 (2017)
44. Iwan S., Małecki K., Korczak J.: Impact of telematics on efficiency of urban freight transport. In: International Conference on Transport Systems Telematics, pp. 50–57. Springer (2013)
45. Ai, N., Zheng, J., Chen, X.: Electric vehicle park-charge-ride programs: a planning framework and case study in Chicago. Transp. Res. Part D **59**, 433–450 (2018)

Planning International Transport Using the Heuristic Algorithm

Mariusz Izdebski[1]([envelope]), Ilona Jacyna-Gołda[2], and Irena Jakowlewa[1]

[1] Faculty of Transport, Warsaw University of Technology, Warsaw, Poland
mizdeb@wt.pw.edu.pl
[2] Faculty of Production Engineering, Warsaw University of Technology,
Warsaw, Poland
jacyna.golda@gmail.com

Abstract. The article refers to the problem of planning international transport. The aim of this paper is to develop the algorithm which will be used to planning international transport. Planning international transport problems are the complex decision problems which refer to the vehicle routing problems and the problems of designating the minimal path in the graph. The approach to planning international transport presented in this paper takes into account the delay times in the intermediate points, e.g. cities and average waiting time at the border crossings. In order to determine the international transport routes the mathematical model was developed, i.e. decision variables, constraints and the criterion function. Decision variables take the binary form and determine the connections between the objects in the transportation network which are realized by the vehicles. Constraints take into account the weight limits on the routes and the time realization of the transportation task. The criterion function determines the minimal transportation route in the context the time of its realization. In order to designate the routes in international transport the heuristic algorithm, i.e. ant algorithm was developed. The steps of building this algorithm was presented. This algorithm was verified in the C# programming language. The results generating by the presented algorithm were compared with the results generating by the random algorithm.

Keywords: Planning international transport · Ant algorithm · Optimization

1 Introduction

The aim of this paper is to develop the algorithm which will be used to planning international transport Planning international transport is a complex decision problem which refers to the transportation company. This complexity arises from the shape of the transportation network and the number of point elements which build this network.

The transportation network is a set of elements (items): suppliers, recipients, etc. which are linked with each other in a diverse way (the direct or indirect relationships between facilities) [1, 2]. Depending on the number of intermediate points on the transport route, the network structure may be single-level [3] (the direct relationship: suppliers – recipients) or multi-level [4, 5] (the indirect relationship: suppliers – intermediate points – recipients) which is called a hierarchical one. One of the

© Springer Nature Switzerland AG 2019
G. Sierpiński (Ed.): TSTP 2018, AISC 844, pp. 229–241, 2019.
https://doi.org/10.1007/978-3-319-99477-2_21

characteristic features of the multi-level network is that cargo has to flow from suppliers via subsequent levels to recipients. In the context of planning international transport the hierarchical transportation network is considered.

Planning international transport problems are the decision problems which refer to the vehicle routing problems [6, 7] and the problems of designating the minimal path in the graph [8, 9]. In the literature these problems were solved by the use of various algorithms, e.g. a genetic algorithm [10, 11], an ant algorithm [12, 13] or dynamic programming [9]. In this paper, planning international transport (transportation routes) was designated by the use of an ant algorithm. Planning international transport by the use of the ant algorithm taking into account the delay time at intermediate points and waiting time at border crossings is the new approach. The authors did not find the application of this approach in planning international transport. Fast time of generating the result by this algorithm is its main feature, what is desired in the process of planning international transport.

This process depends on many factors, e.g. travel times, the size of the tasks. The algorithm for designating this type of problem must be adapted to frequent changes of these factors and generate the solutions in a quick way. In production companies the time of generating the solution by the algorithm plays the most important role. The ant algorithm generates the results in a quick way and therefore this algorithm was selected in this problem.

In the next chapters a mathematical model that determines the routes of vehicles in international transport has been developed, the optimization algorithm for the presented decision problems was discussed.

2   Mathematical Model of the Problem

2.1   Data Input

The transportation network of the company which realizes the international transport consists of the following elements, i.e. the bases, loading points, intermediate points, border crossings, unloading points, Fig. 1.

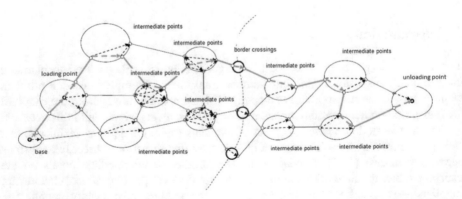

Fig. 1. Transportation network in the international transport (Source: own work)

The main aim is to transport the cargo to the unloading points according to the the the adopted criterion function which is the minimal transportation time of realization the transportation task. The transportation task is defined as loading the cargo form loading points and transport to the unloading place.

The presented mathematical model is intended for designating the international routes of vehicles taking into account waiting times at border crossings. Additionally, delay times in the point elements of the transportation network are taken into account. The return route (abroad – country) in the model is not designated. This route is the same as the route in the relation: country – abroad. The model assumes that between the loading points and the border crossing and between the border crossing and unloading point is at least one intermediate points. The mathematical model of the problem of planning international transport was developed for the following input data:

- V – set of numbers of spot elements of the transportation network: the bases, loading points, intermediate points, border crossings, unloading points $V = \{v: v = 1,2, ..., v', ..., V\}$,
- B – set of number of the bases, $\alpha(v) = \{0, 1, 2, 3, 4, 5\}$ - the mapping which assigns the number of elements of the transportation network to object type, $B = \{v: \alpha(v) = 0 \text{ for } v \in V\}$,
- LP – set of numbers of loading points, $LP = \{v: \alpha(v) = 1 \text{ for } v \in V\}$,
- $IP1$ – set of numbers of intermediate points, e.g. cities (in the country), $IP = \{v: \alpha(v) = 2 \text{ for } v \in V\}$,
- BC – set of numbers of border crossings, $BC = \{v: \alpha(v) = 3 \text{ for } v \in V\}$,
- $IP2$ – set of numbers of intermediate points, e.g. cities (abroad), $IP = \{v: \alpha(v) = 4 \text{ for } v \in V\}$,
- UP – set of numbers of unloading points, $LP = \{v: \alpha(v) = 1 \text{ for } v \in V\}$,
- VOL – set of types of vehicles, $VOL = \{1, ..., vol, ..., VOL\}$,
- $T1$ – time matrix in relations: bases – loading points, $T1 = [t1(v, v'): t1(v, v') \in R^+, v \in B, v' \in LP]$,
- $T2$ – time matrix in relations: loading points – intermediate points (in the country), $T2 = [t2(v, v'): t2(v, v') \in R^+, v \in LP, v' \in IP1]$,
- $T3$ – time matrix in relations: intermediate points (in the country), $T3 = [t3(v, v'): t3(v, v') \in R^+, v \in IP1, v' \in IP1]$,
- $T4$ – time matrix in relations: intermediate points (in the country) - border crossings, $T4 = [t4(v, v'): t4(v, v') \in R^+, v \in IP1, v' \in BC]$,
- $T5$ – time matrix in relations: border crossings - intermediate points (abroad), $T5 = [t5(v, v'): t5(v, v') \in R^+, v \in BC, v' \in IP2]$,
- $T6$ – time matrix in relations: intermediate points (abroad) – intermediate points (abroad), $T6 = [d6(v, v'): d6(v, v') \in R^+, v \in IP2, v' \in IP2]$,
- $T7$ – time matrix in relations: intermediate points (abroad) – unloading points, $T7 = [d7(v, v'): d7(v, v') \in R^+, v \in IP2, v' \in UP]$,
- TL – loading times at loading points, $TL = [tl(v, vol): tl(v, vol) \in R^+, v \in LP, vol \in VOL]$,
- TP – average waiting time at the border crossing, $TP = [tp(v): tp(v) \in R^+, v \in BC]$,
- $TI1$ – delay time at the intermediate points (in the country), $TI1 = [ti1(v): ti1(v) \in R^+, v \in IP1]$,

- **TI2** – delay time at the intermediate points (abroad), **TI2** = $[ti2(v): ti2(v) \in R^+, v \in$ *IP*2],
- **TT** – time of task realization,
- **Q** – the vehicle weight, **Q** = $[q(vol): q(vol) \in R^+, vol \in$ **VOL**],
- **Q1** – the size of cargo which is carried by the vehicle, **Q1** = $[q1(vol): q1(vol) \in R^+,$ *vol* $\in$ **VOL**],
- **L1** – weight limit in relations: loading points – intermediate points (in the country), **L1** = $[l1(v, v'): l1(v, v') \in R^+, v \in LP, v' \in IP1]$,
- **L2** – weight limit in relations: intermediate points (in the country), **L2** = $[l2(v, v'): l2(v, v') \in R^+, v \in IP1, v' \in IP1]$,
- **L3** – weight limit in relations: intermediate points (in the country) - border crossings, **L3** = $[l3(v, v'): l3(v, v') \in R^+, v \in IP1, v' \in BC]$,
- **L4** – weight limit in relations: border crossings - intermediate points (abroad), **L4** = $[l4(v, v'): l4(v, v') \in R^+, v \in BC, v' \in IP2]$,
- **L5** – weight limit in relations: intermediate points (abroad) – intermediate points (abroad), **L5** = $[l5(v, v'): l5(v, v') \in R^+, v \in IP2, v' \in IP2]$,
- **L6** – weight limit in relations: intermediate points (abroad) – unloading points, **L6** = $[l6(v, v'): l6(v, v') \in R^+, v \in IP2, v' \in UP]$.

The type of the decision variables determines the connection between the objects in the transportation network which is realized by the vehicle. The type of the variable formulated as matrix **X1** (relation: bases – loading points), **X2** (relation: loading points – intermediate points), **X3** (relation: intermediate points), **X4** (relation: intermediate points (in the country) - border crossings), **X5** (relation: border crossings - intermediate points (abroad)), **X6** (relation: intermediate points (abroad)), **X7** (relation: intermediate points (abroad) – unloading points) takes the following form:

$$\mathbf{X1} = [x1(v, v', vol): x1(v, v', vol) \in R+, v \in B, v' \in LP, vol \in \mathbf{VOL}] \tag{1}$$

$$\mathbf{X2} = [x2(v, v', vol): x2(v, v', vol) \in R+, v \in LP, v' \in IP1, vol \in \mathbf{VOL}] \tag{2}$$

$$\mathbf{X3} = [x3(v, v', vol): x3(v, v', vol) \in R+, v \in IP1, v' \in IP1, vol \in \mathbf{VOL}] \tag{3}$$

$$\mathbf{X4} = [x4(v, v', vol): x4(v, v', vol) \in R+, v \in IP1, v' \in BC, vol \in \mathbf{VOL}] \tag{4}$$

$$\mathbf{X5} = [x5(v, v', vol): x5(v, v', vol) \in R+, v \in BC, v' \in IP2, vol \in \mathbf{VOL}] \tag{5}$$

$$\mathbf{X6} = [x6(v, v', vol): x6(v, v', vol) \in R+, v \in IP2, v' \in IP2, vol \in \mathbf{VOL}] \tag{6}$$

$$\mathbf{X7} = [x7(v, v', vol): x7(v, v', vol) \in R+, v \in IP2, v' \in UP, vol \in \mathbf{VOL}] \tag{7}$$

2.2 Constraints

The constrains in planning international transport take the form:

- Weight limits on the section of transportation network cannot be exceeded, the sum of the vehicle weight and the cargo weight must be less or equal than the weight limit on a given section of the route, (8) - the relation: loading points - intermediate points, (9) - intermediate points - intermediate points (10) - intermediate points - border crossings, (11) - border crossings - intermediate points. (12) - intermediate points - intermediate points, (13) - intermediate points - unloading points:

$$\forall v \in LP, \forall v' \in IP1, \forall vol \in VOL \quad x2(v,v',vol) \cdot [q(vol)+q1(vol)] \le l2(v,v') \quad (8)$$

$$\forall v \in IP1, \forall v' \in IP1, \forall vol \in VOL \quad x3(v,v',vol) \cdot [q(vol)+q1(vol)] \le l3(v,v') \quad (9)$$

$$\forall v \in IP1, \forall v' \in BC, \forall vol \in VOL \quad x4(v,v',vol) \cdot [q(vol)+q1(vol)] \le l4(v,v') \quad (10)$$

$$\forall v \in BC, \forall v' \in IP2, \forall vol \in VOL \quad x5(v,v',vol) \cdot [q(vol)+q1(vol)] \le l5(v,v') \quad (11)$$

$$\forall v \in IP2, \forall v' \in IP2, \forall vol \in VOL \quad x6(v,v',vol) \cdot [q(vol)+q1(vol)] \le l6(v,v') \quad (12)$$

$$\forall v \in IP2, \forall v' \in UP, \forall vol \in VOL \quad x7(v,v',vol) \cdot [q(vol)+q1(vol)] \le l7(v,v') \quad (13)$$

- The time of task realization must be met, the time of task realization consists of travel time between objects of the network, loading times at loading points, average waiting time at the border crossing and delay time at the intermediate points:

$$\forall vol \in VOL$$

$$\sum_{v \in B} \sum_{v' \in LP} x1(v,v',vol) \cdot [t1(v,v',vol)+tl(v',vol)] + \sum_{v \in LP} \sum_{v' \in IP1} x2(v,v',vol)$$

$$\cdot [t2(v,v',vol)+ti1(v',vol)] + \sum_{v \in IP1} \sum_{v' \in IP1} x3(v,v',vol) \cdot [t3(v,v',vol)+ti1(v',vol)]$$

$$+ \sum_{v \in IP1} \sum_{v' \in BC} x4(v,v',vol) \cdot [t4(v,v',vol)+tp(v',vol)] + \sum_{v \in BC} \sum_{v' \in IP2} x5(v,v',vol)$$

$$\cdot [t5(v,v',vol)+ti2(v',vol)] + \sum_{v \in IP2} \sum_{v' \in IP2} x6(v,v',vol) \cdot [t6(v,v',vol)+ti2(v',vol)]$$

$$+ \sum_{v \in IP2} \sum_{v' \in UP} x7(v,v',vol) \cdot t7(v,v',vol) \le TT$$

$$(14)$$

2.3 The Criterion Function

The criterion function minimizes the total time realization of all tasks. This time depends on travel time between the point elements of the transportation network, average waiting time at the border crossing, delay time at the intermediate points and loading times at loading points. The criterion function takes the following form:

$$\mathbf{F}(\mathbf{X1}, \mathbf{X2}, \mathbf{X3}, \mathbf{X4}, \mathbf{X5}, \mathbf{X6}, \mathbf{X7}) =$$

$$\sum_{v \in B} \sum_{v' \in LP} \sum_{vol \in VOL} x1(v, v', vol) \cdot [t1(v, v', vol) + tl(v', vol)] + \sum_{v \in LP} \sum_{v' \in IP1} \sum_{vol \in VOL} x2(v, v', vol)$$

$$\cdot [t2(v, v', vol) + ti1(v', vol)] + \sum_{v \in IP1} \sum_{v' \in IP1} \sum_{vol \in VOL} x3(v, v', vol) \cdot [t3(v, v', vol) + ti1(v', vol)]$$

$$+ \sum_{v \in IP1} \sum_{v' \in BC} \sum_{vol \in VOL} x4(v, v', vol) \cdot [t4(v, v', vol) + tp(v', vol)] + \sum_{v \in BC} \sum_{v' \in IP2} \sum_{vol \in VOL} x5(v, v', vol)$$

$$\cdot [t5(v, v', vol) + ti2(v', vol)] + \sum_{v \in IP2} \sum_{v' \in IP2} \sum_{vol \in VOL} x6(v, v', vol) \cdot [t6(v, v', vol) + ti2(v', vol)]$$

$$+ \sum_{v \in IP2} \sum_{v' \in UP} \sum_{vol \in VOL} x7(v, v', vol) \cdot t7(v, v', vol) \longrightarrow \min$$

$$(15)$$

3   The Ant Algorithm for Planning International Transport

3.1   Main Assumptions

Theory of ant algorithms introduces the concept of artificial ants. Each ant builds its own route i.e. an ant visits loading point, collects the cargo and transports it via the intermediate points, border crossing to the unloading point.

At first the data input needs to be determined: the set of ants was defined as $MR = \{1, \ldots, mr, \ldots MR\}$, the number of iteration of the algorithm $I = \{1, \ldots, i, \ldots I\}$. The starting point of each route is in the base. The ending point is in the unloading points. The transition probability on the basis which the ant goes to another point in the route takes the following form:

$$PR_{yz}^{mr}(t) = \begin{cases} \dfrac{[\tau_{yz}(t)]^{\alpha} \cdot [\eta_{yz}(t)]^{\beta}}{\sum_{l \in \Omega^{mr}} [\tau_{yl}(t)]^{\alpha} \cdot [\eta_{yl}(t)]^{\beta}}, & z \in \Omega^{mr} \\ 0, & z \notin \Omega^{mr} \end{cases} \qquad (16)$$

$\tau_{yz}(t)$ – the intensity of pheromone trail between the y-the a z-the point in t-the iteration,

$\eta_{yz}(t)$ – the heuristic information, e.g. $\eta_{yz}(t) = 1/w(y, z)$, where $w(y, z)$ - the travel time between y-the point of route and z-the point of the route and the loading time at z-th loading points, or waiting time at the z-th border crossing,

α, β – parameters determining the effect of pheromones and the heuristic information on the behavior of ants,

Ω^{mr} – the set of unvisited points l in the route of the ant.

Random choice of the route between the point y and point l begins from calculating the probability of the transition to the another points according to the pattern (16). The next step is to calculate the distribution for each transition path and draw the number r from the range [0,1]. The route tr about the value of the distribution q_{tr} which fulfils the condition $q_{tr-1} < r \leq q_{tr}$ is selected, where tr is the number of the route between y – the point and l – the point.

After the realization of routes by all ants the pheromone update must be made. Three types of the pheromone update can be distinguished: ant – density, ant – quantity, ant – cycle. In order to update the pheromone the ant – cycle was used as the most efficient version of the ant algorithm [12, 13]. At the beginning it is assumed that the trail on the links between the points is equally strong. In subsequent iterations, the pheromone trail is calculated according to the formula:

$$\tau_{yz}(t+1) = (1 - \rho) \cdot \tau_{yz}(t) + \sum_{mr=1}^{MR} \Delta\tau_{yz}^{mr}(t) \qquad (17)$$

where

mr – another ant $mr \in MR$ $mr \in MR$, ρ – a factor pheromone ($0 < \rho \le 1$), $\tau_{yz}(t+1)$ – the strengthening of the pheromone, for the first iteration this strengthening takes the value τ_0 for each connections, when the route (y, z) was used by mr

$$\Delta\tau_{yz}^{mr}(t) \begin{cases} \frac{1}{L^{mr}(t)} & \text{–this ant} \\ 0 & \text{otherwise} \end{cases} \qquad (18)$$

where:

$L^{mr}(t)$ – the time of the route in t iteration realized by mr – this ant, if the segment of routes (y, z) was realized by mr - this ant then $\Delta \tau_{yz}^{mr}(t)$ equals $1/L^{mr}(t)$, otherwise 0.

3.2 Steps of Algorithm

Steps of the ant algorithm can be presents as follow, Figs. 2 and 3:

- Step 1. Introduce the input data from the mathematical model.
- Step 2. Designate the constraints (14) – only for the relation: the base - loading points. If this constraint is fulfilled for a given relation, the transition probability is calculated from base to each loading point. If the constraint is not fulfilled, the time of task realization must be changed. The ant goes to the loading point from the base on the basis on the calculated probability. This step is repeated until all the loading points are checked (LP – the number of loading points, z-th loading point).
- Step 3. Designate the constraints (8), (14) – only for the relation: the loading point - intermediate points. Additionally, the constrain (14) takes into account the relation with the step 2. If this constraints are fulfilled for a given relation the transition probability is calculated. The ant goes to the intermediate point on the basis on this probability. This step is repeated until all the intermediate points are checked (IP1 – the number of intermediate points, z-th intermediate point).
- Step 4. Designate the constraints (9), (10), (14) – only for the relation: the inter-mediate point - intermediate points and border crossings. Additionally, the constrain (14) takes into account the relation with the step 2, 3. If this constraints are fulfilled for a given relation the transition probability is calculated. The ant goes to the intermediate point or border crossing on the basis on this probability. This step is repeated until all the points are checked.

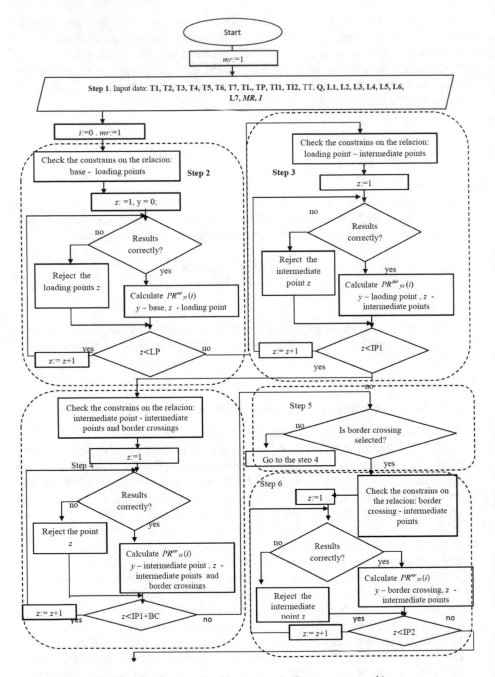

Fig. 2. The ant algorithm – part 1 (Source: own work)

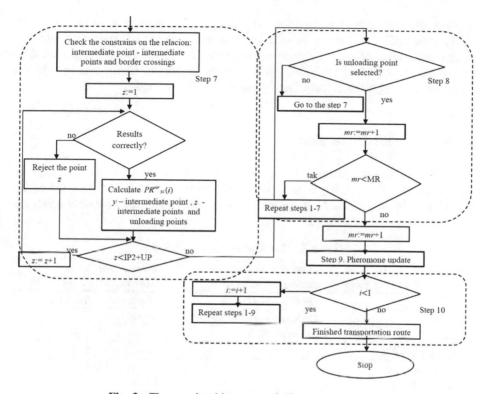

Fig. 3. The ant algorithm – part 2 (Source: own work)

- Step 5. In the case when in the step 4 the border crossing is selected, the step 6 is realized. When in the step 4 the intermediate point is selected, the step 4 is repeated untill the border crossing will be selected.
- Step 6. Designate the constraints (11), (14) – only for the relation: border crossing - intermediate points. Additionally, the constrain (14) takes into account the relation with the step 2, 3, 4. The ant goes to the intermediate point on the basis on calculated probability. This step is repeated until all the intermediate points are checked.
- Step 7. Designate the constraints (12), (13), (14) – only for the relation: intermediate point - intermediate points and unloding points. Additionally, the constrain (14) takes into account the relation with the step 2, 3, 4. 6. The ant goes to the inter-mediate point or unloading points on the basis on this probability. This step is repeated until all the points are checked.
- Step 8. In the case when in the step 7 the unloading points is selected, the ant completes the route. The another ant from the population starts building own route. The steps from 1 to 7 must be repeated until $mr = MR$ (MR – the number of ants in the population). When in the step 7 the intermediate point is selected, the step 7 is repeated untill the unloading point will be selected.
- Step 9. Pheromone update according to the pattern (17).

- Step 10. The number of iterations is checked. The steps from 1 to 9 must be repeated until $i = I$, if the number of iteration is exceeded, the route with the minimum value of the realization time is selected among all the routes generated by the ants in the population. This route is the solution for one transportation task. For the other transportation tasks the algorithm must be run again.

4   Results and Discussion

The algorithm was implemented by the use of the real input data in the C# programming language. The loading point is located in Białystok, loading time – 0.5 h, the unloading point in Brześć, the intermediate points and border crossings were presented

Table 1. Data input for the algorithm.

Intermediate points (country)/delay time [hour]	Border crossings/average waiting time [hour]	Intermediate points (abroad)/delay time [hour]
Białystok/0.5 h	Bobrowniki/2 h	Oberowszczyzna/0.1 h
Kleosin/0.2 h	Połowce - Pieszczatka/6 h	Wysokie/0.2 h
Solniczki/0.1 h	Kukuryki/2 h	Motykały Wielkie/0.1 h
Hryniewicze/0.2 h		Świsłocz/0.2 h
Klewinowo/0.1 h		Porozów/0.1 h
Bielsk Podlaski/1 h		Prużana/0.1 h
Kleszczele/0.2 h		Widomla/0.2 h
Siemiatycze/0.5 h		Czarnawczyce/0.4 h
Łosice/0.4 h		
Międzyrzec Podl./0.4 h		
Gródek/0.3 h		
Konstantynów/0.1 h		
Janów Podlaski/0.4 h		
Dubicze/0.3 h		

in Table 1. The travel times between objects of the network were calculated based on Google Maps, time of task realization – 8 h, the vehicle weight with cargo – 30 tone, the weight limit on the relation: Bielsk Podlaski - Kleszczele – 12 tone.

The first step of implementation of the ant algorithm was to find the set of the best parameters which characterizes this algorithm. The number of iterations was set to 100. The size of population (ants) was set to 200. The combinations of parameters were presented in Table 2. The results of all tests for parameters were presented in Table 3. In order to verify the correctness of the ant algorithm (AA), its results (for the best parameters – test 22) were compared with random values (AL). The ant algorithm in each case generated a better solution than the random algorithm. The results are shown in Table 4.

The solution generated by the ant algorithm for complex decision problems is a sub-optimal solution, which confirms Table 4. However, considering the complexity of

Table 2. The combinations of the parameters for the ant algorithm.

Test	α	β	γ	Test	α	β	γ	Test	α	β	γ
1	1	0.5	0.2	17	1	1	0.2	33	1	5	0.2
2	1	0.5	0.4	18	1	1	0.4	34	1	5	0.4
3	1	0.5	0.6	19	1	1	0.6	35	1	5	0.6
4	1	0.5	0.8	20	1	1	0.8	36	1	5	0.8
5	3	0.5	0.2	21	3	1	0.2	37	3	5	0.2
6	3	0.5	0.4	22	3	1	0.4	38	3	5	0.4
7	3	0.5	0.6	23	3	1	0.6	39	3	5	0.6
8	3	0.5	0.8	24	3	1	0.8	40	3	5	0.8
9	5	0.5	0.2	25	5	1	0.2	41	5	5	0.2
10	5	0.5	0.4	26	5	1	0.4	42	5	5	0.4
11	5	0.5	0.6	27	5	1	0.6	43	5	5	0.6
12	5	0.5	0.8	28	5	1	0.8	44	5	5	0.8
13	10	0.5	0.2	29	10	1	0.2	45	10	5	0.2
14	10	0.5	0.4	30	10	1	0.4	46	10	5	0.4
15	10	0.5	0.6	31	10	1	0.6	47	10	5	0.6
16	10	0.5	0.8	32	10	1	0.8	48	10	5	0.8

Table 3. The results of the ant algorithm for each test.

Test	Results	Test	Results	Test	Results
1	10:25	17	7:35	33	10:24
2	9:45	18	7:05	34	9:35
3	9:20	19	7:15	35	10:45
4	9:35	20	7:00	36	10:23
5	10:03	21	7:20	37	8:40
6	8:30	22	6:35	38	8:00
7	9:05	23	6:56	39	8:25
8	8:50	24	7:45	40	8:34
9	9:25	25	8:00	41	9:02
10	11:20	26	8:45	42	9:34
11	10:24	27	8:30	43	9:09
12	9:53	28	8:20	44	9:23
13	9:48	29	9:24	45	10:23
14	9:26	30	10:20	46	10:11
15	10:04	31	10:25	47	9:27
16	10:37	32	11:10	48	10:29

the planning international transport problem, the solution is accepted from a practical point of view. For many cases the random algorithm did not find the solutions, because the weight limit was exceeded.

Table 4. The comparison of the algorithms.

Test	AA	AL	Test	AA	AL	Test	AA	AL
1	7:30	10:23	17	7:00	8:45	33	7:25	-
2	6:35	9:55	18	6:10	9:30	34	7:10	11:00
3	7:20	11:55	19	6:55	10:15	35	6:35	10:35
4	6:10	-	20	6:10	11:24	36	6:35	-
5	6:35	9:55	21	7:25	-	37	7:00	12:34
6	7:00	11:25	22	7:15	9:45	38	7:25	10:25
7	7:20	12:05	23	6:55	11:45	39	7:25	9:55
8	7:35	11:25	24	6:35	-	40	6:45	12:05
9	7:20	-	25	7:00	12:23	41	7:00	-
10	6:35	11:24	26	6:55	11:23	42	7:15	12:15

5   Conclusion

The aim of the paper is to present the problem of planning international transport taking into account the delay time at intermediate points and waiting time at border crossings. The early convergence to the sub-optimal solution is blocked by the use of the update pheromone ant – cycle in the ant algorithm.

The sub-optimal results generated by the algorithm depends on many factors, e.g.: the combinations of parameters of the algorithm, the number of iterations or ants in population. It should be underlined that results of the ant algorithm depend on the type of the data input which are taken into account in the mathematical model. The presented model does not take into account the random values of the input data. Further research in the context of the presented problem will be conducted taking into account random character of transport process. The developed algorithm can be used in transportation companies. The main advantage of this algorithm is that the results are generated in a quick way, what is very important for these companies. The processes occurring in the transportation companies are the dynamic processes. For this reason, this algorithm must be started a few times depending on the volume of cargo, delay time. In this case, the calculation speed plays a huge role what underlines utility of this algorithm in the transportation companies.

It should be noted that the presented ant algorithm has been used to solve the specific problem of planning international transport. The presented mathematical model of the problem is the original model, not analyzed in the literature, so it is not possible to compare the results generated by other methods with the results obtained by the ant algorithm.

It should be emphasized that the presented algorithm is the starting point for testing other algorithms within the defined research problem. The comparison of random results with the results generated by the proposed ant algorithm emphasized the effectiveness of its operation in the discussed problem. The generated results by the ant algorithm are the basis for further work on the development of new algorithms in the context of the examined problem.

References

1. Jacyna, M., Wasiak, M.: Decision Support in Designing Transport Systems. Tools of Transport Telematics, pp. 11–23. Springer, Berlin (2015)
2. Jacyna, M., Merkisz, J.: Proecological approach to modelling traffic organization in national transport system. Arch. Transp. Pol. Acad. Sci. **30**(2), 32–71 (2014)
3. Sharma, R.R.K., Berry, V.: Developing new formulations and relaxations of single stage capacitated warehouse location problem (SSCWLP): empirical investigation for assessing relative strengths and computational effort. Eur. J. Oper. Res. **177**(2), 803–812 (2007)
4. Geoffrion, A.M., Graves, G.W.: Multicommodity distribution system design by Benders decomposition. Manage. Sci. **20**(5), 822–844 (1974)
5. Sharma, R.R.K.: Modeling a fertilizer distribution system. Eur. J. Oper. Res. **51**(1), 24–34 (1991)
6. Spliet, R., Gabor, A.F.: The time window assignment vehicle routing problems. Transp. Sci. **49**(4), 721–731 (2013)
7. Figliozzi, M.A., Mahmassani, H.S., Jaillet, P.: Pricing in dynamic vehicle routing problem. Transp. Sci. **41**(3), 302–318 (2007)
8. Sharma, D.K., Peer, S.K.: Finding the shortest path in the stochastic networks. Comput. Math Appl. **53**(5), 729–740 (2007)
9. Fan, Y., Kalaba, R., Moore, J.: Shortest paths in stochastic networks with correlated link costs. Comput. Math Appl. **49**(9–10), 1549–1564 (2005)
10. Jacyna-Golda, I., Izdebski, M., Podviezko, A.: Assessment of efficiency of assignment of vehicles to tasks in supply chains: a case study of a municipal company. Transport **32**(3), 243–251 (2017)
11. Lewczuk, K.: The concept of genetic programming in organizing internal transport processes. Arch. Transp. Pol. Acad. Sci. **34**(2), 61–74 (2015)
12. Dorigo, M., Gambardela, L.M.: Ant colonies for the travelling salesman problem. BioSystems **43**(2), 73–81 (1997)
13. Dorigo, M., Gambardela, L.M.: Ant colony system: a cooperative learning approach to the traveling salesman problem. IEEE Trans. Evol. Comput. **1**(1), 53–66 (1997)

Freight Transport Planners as Information Elements in the Last Mile Logistics

Elżbieta Macioszek[✉]

Faculty of Transport, Silesian University of Technology, Katowice, Poland
elzbieta.macioszek@polsl.pl

Abstract. One of the biggest challenges today in the transport of goods on the world is the problem called "last mile logistics". This is a problem concerning the final stage of transport of goods to the customers. The role of information in the last mile logistics acquired from freight transport planners has been presented in this paper. This paper contains a detailed comparison of existing Open Source software solutions for freight transport planning. This paper also provided grounds to perform one of tasks defined under the international S-mile project implemented within the framework of the ERANET Transport III "Sustainable Logistics and Supply Chain" programme analyzing how different technologies should be modified in order to fulfills all the S-mile project objectives. In the paper the results of the analysis to identify the best freight transport planner to use in the S-mile project have been presented.

Keywords: Last mile delivery · Open source software · Logistics
Freight transport · Freight transport planners

1 Introduction

There are many different urban freight transport planning policy on the world. More than any other type of traffic, the urban freight transport is very often touched subject of national, regional and local policies in different policy fields such as economic, environmental planning and also transportation planning. Consequently the policies are different in different countries of the world and have still changed over the time.

The last mile delivery - according definition [1–4] there is a term used in transportation planning and supply chain management in order to describe the movement of goods and people from a transportation hub to a final destination (f.ex. from home or work). Transporting goods via freight rail networks and container ships is often the most efficient and cost-effective manner of shipping. However, when goods arrive at the high-capacity freight station or port, then they must be transported to their final destination. This last leg of the supply chain is often less efficient comprising up to 28% of the total cost to move goods. Last mile delivery problem can also include the challenge of making deliveries in urban areas where restaurants, retail stores and also other merchants in a central business district often contribute to safety problems and primarily congestion.

The last mile delivery in cities is not only a huge logistic problem but also an important city planning issue. According Pulikottil [5] in the next few years the last

© Springer Nature Switzerland AG 2019
G. Sierpiński (Ed.): TSTP 2018, AISC 844, pp. 242–251, 2019.
https://doi.org/10.1007/978-3-319-99477-2_22

mile deliveries are expected to increase as a result of escalated online trade and also the e - commerce sale will increase on the world wide by 20% and reach an approximate 1.5 trillion dollars. Changes in demand for global products and the increased complexity of logistics and supply chain networks are also expected. Furthermore, there are request for a greater needs of goods by consumers, with the reduction the life cycle of products and limited capacity of warehouses sales floor. The increased demand and the reduction in warehouse capacity causes increased last mile demand frequency in business to business last mile deliveries (B2B) and business to consumer last mile deliveries (B2C). The business to business includes distributors and retailers, parts, suppliers of groceries and large items (f. ex. furniture, devices and machines). According to different international literature (f.ex. Ehmike [6]) first and last mile deliveries have about share of 10% in the total urban transport. Moreover as much as about 90% of last mile delivery are made by road. The last mile delivery involve transportation over short distances with smaller trucks. Except, in places where the urban infrastructure enables deliveries via water or via rail the first and last mile deliveries would also be possible through the suitable water canals of urban rail terminals.

This paper contains a detailed comparison of existing Open Source software solutions for freight transport planning. The role of information in the last mile logistics acquired from freight transport planners has been presented in this paper. This paper also provided grounds to perform one of tasks defined under the international S-mile project implemented within the framework of the ERANET Transport III "Sustainable Logistics and Supply Chain" programme analyzing how different technologies should be modified in order to fulfills all the S-mile project objectives. In the paper the results of the analysis to identify the best freight transport planner to use in the S-mile project have been presented.

2   Freight Transport Planners as Information Elements in the Last Mile Logistics

The freight transport planners are one of the solutions of Intelligent Transport Systems (ITS) used in freight transport management (f. ex.: [7–17]). Freight transport planners can also plays a role of information elements in the last and first mile logistics because they allows visualizing, choosing and comparing possible routes, as well as supports their implementation in a given transport network between different delivery points. Freight transport planner should reduce the time take to plan to transportation schedule and not only do route planning systems lower mileage, it also help improve asset utilization, cut fuel usage, decrease carbon emissions and increase customer service. The freight transport planners should offers the following capabilities [18]:

- enhances fixed routes and schedules, calculating optimized routes and schedules while meeting required customer delivery windows, truck capacities, driver hours, and other transportation restrictions in order to manage a transportation operation with regular order dates and quantities,

- automatically calculating efficient truck routes and multi-stop schedules which will reduce overall distances, fleet costs, and daily planning efforts,
- optimizes deliveries continually. When a new orders are added, system should continually re-optimizes schedules will maximize efficiency by taking into account delivery areas, available resources, and existing deliveries already confirmed,
- links with live vehicle tracking. Live vehicle tracking allows managers to detect anomalies in route times and distances so they can act immediately to control costs. Comparing planned to actual routes ensures drivers are following the plan. If any deviation occurs, customers should be alerted to delays,
- embedded considers "what-if" scenarios. Using historic data to prepare for vehicle size changes, shifting driver hours, and alternative delivery locations for distribution networks will improve transport efficiency,
- uses multi-period planning. Multi-period planning should decides about the best delivery patterns for each customer, ensuring that multiple deliveries to the same customer are sufficiently spread out across the planning period, while also combining deliveries geographically and balancing workload across the period. Allocating delivery profiles in this way ensures customer delivery requirements, while also minimizing transportation costs,
- combines central scheduling. Combining central scheduling of all fleet movements gives transportation planners the ability to plan nationally or regionally. Inter-depot trucking movements, supplier collections, and packaging disposal can be incorporated to drastically reduce costs and create significant efficiencies,
- reporting. Key performance indicators and business intelligence reporting allows companies to detect operational trends, predict cost implications, and identify possible preventive measures.

Moreover, the freight transportation should provide the most efficient use of transport system. Some basic functionalities of the freight transport planners include [19–21]:

- module converting names to specific locations (converting array of characters written by user into geographical coordinates of specific buildings or streets),
- routing module, namely algorithm (or set of algorithms) setting optimized route between the different points of supplies goods according to required criteria and available transport system resources and transportation companies (including dynamic and selective routing),
- module converting information about route (obtained from routing) into messages understood by user (f. ex. drive straight, in 530 m turn right, etc.).

The freight transport planners have different possibilities of classifying routes by defining one or several criteria. Most frequently the truck drivers makes a choice between the shortest and the fastest route. However there are also more complex solutions. For example the freight transport planners can also define the most environmentally friendly routes.

3 Selecting Freight Transport Planner

A freight transport planner is a system solution (developed in different technologies) which allow to calculated the best path for delivery vehicle between two or more locations in the transport network. A freight transport planner enables selecting a route consisting of several - more or less it depend on the number of delivery points - stretches between different delivery points with the use one mean of transport (usually truck). The planner platform is a programming environment it means set of programming tools with planner modules, data about transport service companies, data about the transport network and servers that enable remote implementation of the planner for various transport networks. The freight transport planner is usually used in a specific transport network and its implementation is unique and local.

One of tasks of the S-mile Project [22–24] was to analysis the existing Open Source planners and choosing the most appropriate one for freight transport implementing. This one will be developed during the S-mile project. Below are presented findings of this comparison involving several out of several hundred analyzed planners.

In a first step of the analysis sixteen out of several hundred analyzed planners were selected. They are Open Source planners as well as not having Open Source features, but with features interesting from the point of view of the objectives of the S-mile project like: vehicle speed, information about the price for a transport, interesting interface, clear structure and others features. The results of preliminary freight transport planners segregation have been presented in Table 1.

Table 1. The results of preliminary freight transport planners segregation.

No.	Planner name	Source
1.	My route online route planner	http://www.myrouteonline.com
2.	Freight journey planner	http://www.freightjourneyplanner.co.uk
3.	Graphhopper	http://www.graphhopper.com
4.	Mapbox	http://www.mapbox.com/industries/logistics
5.	Open trip planner	http://www.opentripplanner.org/
6.	Mobycon	http://www.mobycon.com/action/news/item/980/Free/
7.	Driving route planner	http://www.drivingrouteplanner.com
8.	Borusan Guvencesiyle	http://www.etasimacilik.com
9.	Truck and driver	http://www.truckanddriver.co.uk/free-truck-route-planner/
10.	Loginext	http://www.loginextsolutions.com
11.	Skobbler	http://maps.skobbler.com/
12.	Routinio	http://www.routino.org/
13.	Your navigation routing service	http://yournavigation.org/
14.	Open route service	http://openrouteservice.org/
15.	Open Source routing machine	http://project-osrm.org/
16.	Reit Und Wanderkarte	http://www.wanderreitkarte.de/

To be generally available in any transport network, the planner should be pro-grammable as Open Source [25]. Such planner should be based on a platform which enables several options and modifications adjusting it to conditions in the local transport network and specific requirements. Possible requirements can include f. ex. ability to avoid congested routes. This applies to the layer of data, API (Application Programming Interface) and routing algorithms used in the planner. On the world, a large number of planners is used but only some of them meet conditions specified above. The majority of planners are commercial (like f.ex. Contus Dart Renders [26]) and implemented for a specific area of the transport network or selected transport companies (like f.ex.: Turkish planner - Borusan Guvencesiyle [27]). Such planners have been rejected due to the objectives of the S-mile project, since the project focuses on building a planner which is open and generally available.

Beside, the project looked for planners which can be considered complete and comprehensive - platforms integrating all tools necessary for building and imple-menting it in a given transport network. It means that a single software package should integrate all available functionalities of large transport networks. The planner should also be flexible and modifiable.

After research among existing platforms and planners, there were selected nine out of hundreds of Open Source or partially free products (e.g. restrictions applied to side of transport networks maps). The list of selected planners defined as Open Source is presented in Table 2 (in some cases they may include paid services).

Table 2. Open Source planners for further detailed analysis.

No.	Planner name	Source
1.	Graphhopper	http://www.graphhopper.com
2.	Mapbox	http://www.mapbox.com/industries/logistics
3.	Open trip planner	http://www.opentripplanner.org/
4.	Skobbler	http://maps.skobbler.com/
5.	Routinio	http://www.routino.org/
6.	Your navigation routing service	http://yournavigation.org/
7.	Open route service	http://openrouteservice.org/
8.	Open Source routing machine	http://project-osrm.org/
9.	Reit Und Wanderkarte	http://www.wanderreitkarte.de/

Furthermore, also another criterion for analysis was the availability in public domain of information about a transport network for the planner, technical documen-tation and source code. The majority of planners used such data formats like OSM (Open Street Map) and GTFS (General Transit Feed Specification), so the criterion was met by many of planners.

A planner which is the most appropriate for further modification needs to meet also other criteria - e.g. free and available source code. This requires web based resources to archive the planner's code, which needs to be free, complete and written in one of the popular programming languages. Unfortunately, despite what was initially declared by planners developers, this criterion became a barrier for many of them. After examining repositories documenting of each planner, the number of planners was limited up to eight. Table 3 presents a list of planners:

- capable of using public data about transport network,
- open and having public source code for routing algorithms,
- having public repository containing complete platform, and
- provided with technical documentation.

Planners presented in Table 3 were selected as potential substitution to Open Trip Planner platform since they are complete, full openness, have multiple implementation, full accessibility and functionalities. In turn, Table 2 presents all potential free platforms, however some of them have limited licenses in terms of data transfer or selected components. It means that platforms may have only partial application in open platforms.

Table 3. Selected complete repository and public (not necessarily free) Open Source planner platforms.

No.	Planner name	Source
1.	Graphhopper	http://www.graphhopper.com
2.	Mapbox	http://www.mapbox.com/industries/logistics
3.	Open trip planner	http://www.opentripplanner.org/
4.	Skobbler	http://maps.skobbler.com/
5.	Routinio	http://www.routino.org/
6.	Your navigation routing service	http://yournavigation.org/
7.	Open Source routing machine	http://project-osrm.org/
8.	Reit Und Wanderkarte	http://www.wanderreitkarte.de/

The level of sources for planners presented in Table 3 varied as regards their quality and capacity. There are significant differences between planners which comply with the above criteria from the point of view of their future use. There are such differences as:

- programming language used in source code (in fact selection of such languages is used),
- different routing options (number of routing algorithms available and implemented functionalities),
- environment (operating system).

The programming language is not a strict criterion for selecting a planner. It is just a technical issue. The majority of planners is based on solutions integrating several programming languages at the same time, such as C++, Java (including scripts), Phyton, PHP and other. Different languages are used for different functionalities. As regards routing algorithms C++ and Java as well as other languages can be used. The least important layers is the API layer which is a mere graphic interface.

After the analysis of the accessibility of planners presented in Tables 1, 2 and 3 there can be taking into consideration only one trip planner, namely the Open Trip Planner [28]. The quality of these planner platform has been confirmed by its implementation in different countries on the world in many planners f.ex. in:

- TriMet [29]. Popular trip planner used f.ex. in Portland, USA,
- Green Travelling [30, 31]. Universal multimodal trip planner which can be used in any region on the world (after implementation of the revelant data characterize this area like f.ex.: map of the area, timetables for public transport etc.) This planner has been tested with a positive results in Turkey, Spain and in Poland.

The Open Trip Planner platform meets almost all project criteria. The Open Trip Planner is based on open data code for freight transport planning (General Transit Feed Specification and Open Street Maps). It has source codes for routing algorithms (A* and Dijkstra [32, 33]). It also has a special routing module for specific RAPTOR networks. The repository [34] contains source codes and documentation. The documentation in the GitHub repository includes development stages of the planner beginning with the oldest versions. The planner also has a WebAPI layer implemented. Initial tests of the Open Trip Planner platform for a dense transport network in the Upper Silesia Agglomeration in Poland were successful. This result providing a good basis for further developing a freight transport planner.

Several another planners analyzed in the S-mile project can be quickly adjusted to requirements of a generally available and open freight transport planner. Such planners are presented in Table 4. Comments in Table 4 regarding existing planners provide a subjective overview.

Table 4. Selected open source planner platforms quickly adjustable to open trip planner platform functionality (advantages and disadvantages).

No.	Planner name	Remarks
1.	Graphhopper	Good documentation, low evaluation versions (development platform), wide selection of platforms
2.	Mapbox	Good documentation, stable platform
3.	Skobbler	Limited number of platforms (Android, iOS), more for navigation than for planner platform
4.	Routinio	Stable platform, no service for mobile devices, several development versions
5.	Your navigation routing service	No GTFS service, no service for mobile devices, problem with update
6.	Open route service	No GTFS service, incomplete Open Source
7.	Open Source routing machine	No GTFS service, no service for mobile devices
8.	Reit Und Wanderkarte	Example of specialized platform, based on using external sources

4 Conclusions

The comparison of planners presented in this paper shows that despite apparently large number of planner platforms only limited number of them is fully available to the public. This applies not only to the licensing policy, but also physical accessibility of resources in repositories of specific planners. Actually only one planner bodes well for building a accessible planner.

The analysis presented in this report allowed to indicate that the Open Trip Planner is the planner most flexible and susceptible for further modification in the project. However, in order to adapt to the requirements contained in the project, numerous modifications and different works with Open Trip Platform are needed, including among other:

- changes in interface,
- changes in structure,
- extension of the search modes because of the different criteria (f.ex. due to the reduction of pollutant emissions),
- and other.

Acknowledgements. The present research has been financed from the means of the National Centre for Research and Development as a part of the international project within the scope of ERA-NET Transport III Programme "Smart platform to integrate different freight transport means, manage and foster first and last mile in supply chains (S-MILE)".

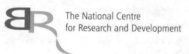

References

1. Brigitte, A.: Improving freight efficiency within the 'last mile': a case study of Wellington's central business district. Thesis for Master of Planning. University of Otago, Otago (2012)
2. Rodrigue, J.P., Comtois, C., Slack, B.: The Geography of Transport Systems, 2nd edn. Routledge, London (2009)
3. Scott, M., Anderka, S., O'Donnell, E.: Improving Freight Movement in Delaware Central Business Districts. Institute for Public Administration, University of Delaware, Delware (2009)
4. Macioszek, E.: First and last mile delivery-problems and issues. In: Sierpiński, G. (ed.) Advanced Solutions of Transport Systems for Growing Mobility. AISC, vol. 631, pp. 147–154. Springer, Cham (2018)
5. Pulikottil, G.P.: E-commerce-the last mile. http://www.linkedin.com/pulse/e-commerce-last-mile-p-graceson-paul
6. Ehmke, J.: Integration of Information and Optimization Models for Routing in City Logistics. International Series in Operations Research & Management Science, vol. 177. Springer, New York (2012)

7. 2016 Logistics Planner. Inbound Logistics. Special Supplement, http://www. inboundlogistics.com
8. Małecki, K., Iwan, S., Kijewska, K.: Influence of intelligent transportation systems on reduction of the environment negative impact of urban freight transport based on szczecin example. Proc. Soc. Behav. Sci. **151**, 215–229 (2014)
9. Regan, A., Holguin-Vcras, J., Chow, G., Sonstegaard, M.: Freight transportation planning and logistics. http://onlinepubs.trb.org/onlinepubs/millennium/00044.pdf
10. Tseng, Y., Yue, W.L., Aptaylor, M.: The role of transportation in logistics chain. Proc. East. Asia Soc. Transp. Stud. **5**, 1657–1672 (2005)
11. Macioszek, E.: The application of HCM 2010 in the determination of capacity of traffic lanes at turbo roundabout entries. Transp. Probl. **11**(3), 77–89 (2016)
12. Wątróbski, J., Małecki, K., Kijewska, K., Iwan, S., Karczmarczyk, A., Thompson, R.: Multi-criteria analysis of electric vans for city logistics. Sustainability **9**(8), 1453, 1–34 (2017)
13. Małecki, K.: Graph cellular automata with relation-based neighbourhoods of cells for complex systems modelling: a case of traffic simulation. Symmetry **9**(12), 322, 1–22 (2017)
14. Pypno, C., Sierpiński, G.: Automated large capacity multi-story garage-concept and modeling of client service processes. Autom. Constr. **81C**, 422–433 (2017)
15. Turoń, K., Czech, P., Juzek, M.: The concept of walkable city as an alternative form of urban mobility. Sci. J. Silesian Univ. Technol. Ser. Transp. **95**, 223–230 (2017)
16. Turoń, K., Golba, D., Czech, P.: The analysis of progress CSR good practices areas in logistic companies based on reports "Responsible Business in Poland. Good Practices" in 2010–2014. Sci. J. Silesian Univ. Technol. Ser. Transp. **89**, 163–171 (2015)
17. Golba, D., Turoń, K., Czech, P.: Diversity as an opportunity and challenge of modern organizations in TSL Area. Sci. J. Silesian Univ. Technol. Ser. Transp. **90**, 63–69 (2016)
18. Salter, W.: Choosing a Route Planning System. Inbound Logistics, http://www. inboundlogistics.com/cms/article/choosing-a-route-planning-system
19. Celiński, I.: Using GT planner to improve the functioning of public transport. In: Macioszek, E., Sierpiński, G. (eds.) Recent Advances in Traffic Engineering for Transport Networks and Systems. LNNS, vol. 21, pp. 151–160. Springer, Cham (2018)
20. Celiński, I.: Transport network parametrisation using the GTAlg tool. In: Macioszek, E., Sierpiński, G. (eds.) Contemporary Challenges of Transport Systems and Traffic Engineering. LNNS, vol. 2, pp. 111–1231. Springer, Cham (2017)
21. Celiński, I.: Support for green logistics using the GTAlg tool. In: Sierpiński, G. (ed.) Intelligent Transport Systems and Travel Behaviour. AISC, vol. 505, pp. 121–134. Springer, Cham (2017)
22. Staniek, M.: Stereo vision method application to road inspection. Balt. J. Road Bridge Eng. **12**(1), 38–47 (2017)
23. Staniek, M.: Road pavement condition as a determinant of travelling comfort. In: Sierpiński, G. (ed.) Intelligent Transport Systems and Travel Behaviour. AISC, vol. 505, pp. 99–107. Springer, Cham (2017)
24. Staniek, M.: Moulding of travelling behaviour patterns entailing the condition of road infrastructure. In: Macioszek, E., Sierpiński, G. (eds.) Contemporary Challenges of Transport Systems and Traffic Engineering. LNNS, vol. 2, pp. 181–191. Springer, Cham (2017)
25. Open Source Initiative. https://opensource.org/osd
26. Contus Dart: Last Mile Delivery Tracking Software. http://www.contus.com/on-demand-delivery-app.php
27. Borusan Guvencesiyle. http://www.etasimacilik.com
28. Open Trip Planner. http://www.opentripplanner.org
29. Trimet. http://trimet.org

30. Sierpiński, G.: Distance and frequency of travels made with selected means of transport: a case study for the Upper Silesian Conurbation (Poland). In: Sierpiński, G. (ed.) Intelligent Transport Systems and Travel Behaviour. AISC, vol. 505, pp. 75–85. Springer, Cham (2017)
31. Sierpiński, G.: Technologically advanced and responsible travel planning assisted by GT planner. In: Macioszek, E., Sierpiński, G. (eds.) Contemporary Challenges of Transport Systems and Traffic Engineering. LNNS, vol. 2, pp. 65–77. Springer, Cham (2017)
32. Brodesser, M.: Multi-Modal Route Planning. University of Freiburg, Freiburg (2013)
33. Dijkstra, E.W.: A note on two problems in connection with graphs. Num. Math. **1**, 269–271 (1959)
34. Open Trip Planner Repository. https://github.com/opentripplanner

The Competitiveness of Inland Shipping in Serving the Hinterland of the Seaports: A Case Study of the Oder Waterway and the Szczecin-Świnoujście Port Complex

Izabela Kotowska[1]([✉]), Marta Mańkowska[2], and Michał Pluciński[2]

[1] Faculty of Engineering and Economics of Transport,
Maritime University of Szczecin, Szczecin, Poland
i.kotowska@am.szczecin.pl
[2] Faculty of Management and Economics of Services, University of Szczecin,
Szczecin, Poland
{marta.mankowska,michal.plucinski}@wzieu.pl

Abstract. The aim of this article is to verify the existing knowledge in the area of competitiveness factors of hinterland modes of transport in land-sea transport chains. In order to fulfill the primary objectives of the article, a case study of the Oder Waterway and Szczecin-Świnoujście port complex was applied. The verification of competitiveness factors of inland shipping in serving seaports hinterland was made on the basis of in-depth interviews among shippers who represented the most important exporters and importers using sea transport, and whose premises were located in the areas that gravitate towards inland waterway transport. The study provided evidence that the most important factor is to assure reliability of delivery. It is the necessary condition for a modal shift between competitive transport branches. Lower transport costs without the certainty of delivery on time will not convince shippers to a modal shift.

Keywords: Inland shipping · Hinterland transport
Transport competitiveness factors · Oder waterway
Szczecin-Świnoujście port complex

1 Introduction

The quality of the hinterland transport system is one of the main factors affecting seaport competitiveness [1–3]. Land transport, and road transport in particular, is still predominant in serving the seaports. Nevertheless, an increase in the volume of port traffic, which is now being observed, requires even more efficient cargo handling in the hinterland. Therefore, in the regions with extensive networks of inland waterways, including Western Europe, a real alternative to road and rail transport is intermodal transport in the form of river-land transport chains [4–11].

The importance of inland shipping as a significant transport link to the hinterland is more and more realized by the Baltic seaports authorities. An example is the Szczecin-Świnoujście port complex, which comprises two out of four ports of primary importance to the Polish economy (in addition to the ports of Gdańsk and Gdynia), with

© Springer Nature Switzerland AG 2019
G. Sierpiński (Ed.): TSTP 2018, AISC 844, pp. 252–263, 2019.
https://doi.org/10.1007/978-3-319-99477-2_23

annual cargo handling volume of over 20 m tonnes. The Szczecin-Świnoujście port complex has access to the Oder Waterway (OW, E-30) which is the most important waterway in Poland and an important link in the Baltic-Adriatic transport corridor. The course of the Oder River in Poland is presented in Fig. 1.

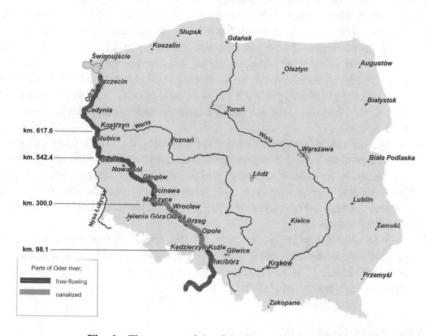

Fig. 1. The course of the Oder River (Source: [12])

However, due to the long-standing degradation of the hydraulic structures and lack of modernization, its operational parameters do not enable regular navigation (Table 1).

Table 1. The parameters of the Oder River (Source: [13])

Sections of the Oder River	Length (km)	Class
Racibórz – Kędzierzyn Koźle	44.4	Ia
Kędzierzyn Koźle – Brzeg Dolny	187.1	III
Brzeg Dolny – Nysa Łużycka estuary	259.8	II
Nysa Łużycka estuary – Warta estuary	75.2	II
Warta estuary – Ognica	79.4	III
Ognica – Dąbie lake	44.6	Vb

In consequence Oder Waterway is currently of very little importance in serving the Szczecin-Świnoujście port complex. Although its share amounted to 14%–16% in the 1980s, afterwards it decreased steadily to fall to merely 5% in 2015 [14]. Many research studies have proven that the quality of an inland waterway network is the primary factor affecting development of inland shipping, which contributes to reducing the barge transport costs, improving the quality of the services and increasing the speed of delivery [15–17]. Stimulation of inland shipping to/from the Szczecin-Świnoujście port complex primarily requires upgrading of the OW operational parameters.

After many years of negligence, restoration of navigability of inland waterways in Poland was included in the list of key projects in the government investment plans with regard to the development of transport infrastructure from 2016 to 2030. The planned investment measures aim at adapting the Polish inland waterways to the requirements of shipping routes of international importance (AGN agreement). The investment measures to make the OW navigable are of the highest priority [18] and are designed to eventually upgrade this waterway to class Va (which ensures operation of push barges of up to 3 k tonnes).

Primarily, this article attempts to answer the following questions: will the improved quality of the inland waterway infrastructure contribute to shifting the demand from rail and road transport to inland shipping and contribute to inter-corridor shipment (switching between transport corridors)? What factors (besides the infrastructure quality) are responsible for competitiveness of river-land transport chains in serving the hinterland of the seaports?

In order to fulfil the primary objectives of this article, a case study of the Oder Waterway and Szczecin-Świnoujście port complex was applied. The verification of competitiveness factors of inland shipping was made on the basis of demand survey (in-depth interviews among shippers).

2 Literature Overview

The literature examines a number of transport competitiveness factors, however, most researchers consider only a few of them as important, without hierarchizing their importance. For example, Cullinane and Toy [19] indicated a list of 15 categories of modal choice criteria, but the research results showed that the most important factors were: transit time reliability, speed, and transport cost. Moon et al. [20] indicated quantitative factors such as: distance, time and cost, and qualitative factors: reliability, flexibility, frequency, freight information service, safety, transport route/mode awareness.

The majority of researchers underline that the cost of transport is one of the most important criteria in modal, followed by loss and damage, transit time, flexibility, late arrivals, and frequency choice, e.g. [21–26]. Wiegmans and Konings [27], claim that quality aspects such as reliability, safety, security, etc. also play a certain role, but to a lesser extent than costs. They investigated the influence of the economies of scale in inland waterway transport and terminal operations.

An increasing number of studies analyse the influence of transport costs on the scale of modal shift, e.g. [28, 29]. However most of them focus on direct comparison of costs between transport chains rather than shippers' tendency to a modal shift to combined transport [30, 31]. In practice, shippers are inclined to make changes to their transport chains and switch to another transport mode only when the price difference is at a sufficient level, otherwise they stick to their tested and proven transport solutions. The research done by Frémont and Franc [6] proved that, from shippers' point of view, a modal shift is possible when prices of combined transport are between 10% and 20% lower than prices of road transport. There is no similar research concerning conventional (non-combined) land-sea transport chains.

Only a few studies have shown that the importance of transport competitiveness indicators also depends on external factors. Bergantino et al. [32] proved that the importance of competitive factors depends on the company size (the greater the company size, the lower the sensitivity to time), size of the cargo consignment (the greater the load factor, the higher the importance of punctuality), size of the operator (larger operators assign higher rating to the risk of loss/damage). Wiegmans and Konings [27] point that the decisions are also determined by the type of cargo. The cost of transport is the most important in the case of bulk cargo, whereas in the case of containers, the most important factors are: costs, transport time, reliability and frequency of the service.

The verification of competitiveness factors of inland shipping in serving seaports hinterland (indicated in the literature) is the main object of the research undertaken in this article.

3 Methodology

The verification of competitiveness factors of inland shipping in serving Szczecin and Świnoujście seaports hinterland required detailed demand surveys. The study process is shown in Fig. 2. Completion of the subsequent stages of the study was based on the primary and secondary sources. In the first stage, based on the data provided by the Analytical Centre of the Customs Administration (ACCA), the analysis was made with regard to the volume and structure of trade conducted by the entities registered in the hinterland of the ports of Szczecin and Świnoujście, located in the inland shipping catchment area (ACCA, as of 2014). 5 Polish provinces located along the OW, i.e. Lubuskie, Wielkopolskie, Dolnośląskie, Opolskie, and Śląskie, were included in the study. In the next step, the area covered by the study was narrowed down to the land located within 50 km from the OW.

To identify and prioritise factors determining the transport competitiveness in the seaports hinterland and to assess the level and the structure of demand for transport to be provided by inland shipping to/from the seaports of Szczecin and Świnoujście, individual in-depth interviews were carried out.

The next step was to find interviewees – these were selected from the actors of the land-sea transport chains. The first group was made up by shippers identified in the ACCA data, who represented the most important exporters and importers using sea transport, and whose premises were located in the areas that gravitate towards inland waterway transport. The list of maritime importers and exporters is based on the factual data from the SAD and INTRASTAT documents, and it was acquired from the 2014 official records provided by the ACCA.

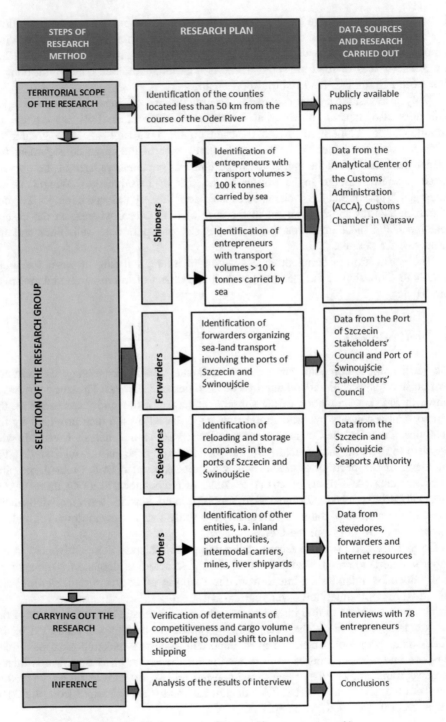

Fig. 2. Study process diagram (Source: own work)

In order to complement the research group of shippers, interviews were also conducted among the largest stevedoring companies in Szczecin and Świnoujście seaport complex.

Finally, that group of respondents consisted of:

- 11 (6 exporters, 5 importers) representatives of entities who export or import over 100 k tonnes of cargo per year.
- 32 (13 exporters and 19 importers) representatives of entities who generate operations at the level of 10 k–100 k tonnes per year.

This group included mainly producers and traders of such products as coal and other mineral products, metals, chemicals, timber, paperboard and other miscellaneous manufactured articles.

The interviewees were people responsible for the organization of logistics at different levels of management. Depending on the organizational structure of the individual company, these were: directors/chief executives, development directors, logistics directors, freight forwarding directors, sales and marketing directors, rail and maritime transport managers, global buyers, logistics coordinators and other managers. The selection of the respondents was determined by their knowledge and experience in the field of transport processes implemented in their companies. The choice of these respondents has also been a consequence of the applied research method (in-depth interviews).

The second group of respondents comprised forwarders organizing transport through the seaports in Szczecin and Świnoujście. This group included 13 companies cooperating with operators of terminals located in Szczecin and Świnoujście seaports. These forwarders were mostly the members of the Port of Szczecin Stakeholders' Council.

The third group of respondents involved 5 stevedoring enterprises which handle transshipment operations at the seaports of Szczecin and Świnoujście.

The fourth group included 17 other respondents: inland port authorities, intermodal carriers, mines, river shipyards, power plants and cogeneration plants as well as business entities located in the OW hinterland who had not participated in maritime trade yet, but were interested in development of their business in that field. The selected group of entities were in-depth interviewed in the third quarter of 2015. All of the interviewees were asked the three questions:

1. Do you use inland shipping in cargo transport from/to seaports? If not, why?
2. If the OW were navigable on 250 days per year, what factors would be responsible for shifting cargo from road or rail transport to inland shipping?
3. Assuming that the expected demand for transport is met, what volume of cargo would you be willing to transport?

At the same time, conclusions drawn from the direct interviews induced the authors to conduct supplementary analysis to see how inland shipping is cost competitive with rail and road transport in serving the seaports hinterland.

4 Results: The Determinants of Intersectoral and Inter-Corridor Shifts of Demand

The survey carried out on the group of shippers described in the methodology part showed a great potential for the modal shift to inland shipping.

In reply to Question 1: *"Do you use inland shipping in cargo transport from/to seaports? If not, why?"* almost all the interviewees stated that they made no use of inland shipping because **the reliability of delivery was not assured**. The reliability of delivery is defined as the probability of satisfying the announced delivery time [33]. No inland carrier is able to assure continuous deliveries because of varying water levels. Therefore, they use rail transport with supplementary road transport. Inland shipping was used by sand mines, but only in local transport, and by inland shipyards which sporadically transported newly-built inland vessels to seaports. In this case the reliability of delivery means that OW will be navigable on 250 days per year.

However the study carried out on the stevedoring companies in Szczecin and Świnoujście showed that they had experience in handling barge transport provided within the East to West rail-barge chains, i.e. Poland – Western Europe (mainly Germany), with using Szczecin-Berlin inland waterway, which has better operational parameters (minimum class III). Cargo such as coal, scrap metal or fertilizers is transported in this manner. The stevedores would also like to use the above mentioned experience in transport provided in the north-south direction along the OW to/from the main supply base of the Szczecin – Świnoujście port complex.

Answering Question 2: *"If the OW were navigable on 250 days per year (the reliability of delivery will be ensured), what factors would be responsible for shifting cargo from road or rail transport to inland shipping?"* Apart from the reliability of delivery, in the opinion of the surveyed shippers of cargo as well as of the other respondents, the main factor in choosing the means of transport is **the cost of transport**. A sufficient condition for shifting their cargo from rail/road transport towards inland shipping is just a 10% reduction in the freight rate. Most of the surveyed shippers have no direct access to the OW. Therefore, a reduction in the freight costs would have to relate to the cost of the whole river-land chain, i.e. inland shipping, port transloading and pre-haulage transport fee. In case an additional handling operation in an inland port caused a partial damage of cargo, the reduction of the price would have to be at least 20%.

All interviewed cargo shippers and other entities would also accept a longer delivery time, if it were compensated by a reduction in the overall transport costs. However, the key issue for them is appropriate arrangement of deliveries.

Apart from the cost and duration of transport, other factors also have their influence on the modal shift to inland shipping. Their importance is in connection with the specificity of services provided for a particular subsection of cargo, including:

a. transport security – for example in the case of liquid chemicals the shippers consider inland shipping to be safer and more reliable than rail transport;
b. size of a single consignment – in the case of consignments smaller than a full train load, the cost of rail transport rises and it may have an impact on the modal shift to inland shipping.

The containerized cargo shippers who export goods to overseas markets were more willing to change their shipment logistics. They see a greater potential in the river-land chains providing that the costs of transport are reduced by at least 10%–20% and a logistic operator takes over organisation of the supply chain. Such a group includes global players in the chemical sector who export cargo in containers to their sub-sidiaries in other countries, and importers of steel semi-products used for their own production. Competitiveness of river-land transport chains vs. unimodal rail or road transport is also determined by appropriate storage potential of inland ports. A good example is granite shipment logistics, in which cost of transport may exceed 50% of the total cost of a sale transaction. As a result, sale transactions are conducted in the very seaport or port city and shipment to the hinterland is arranged by the final client. With development of barge transport it would be possible to shift distribution of this cargo to the hinterland.

The location of production facilities is another important factor which affects the choice of transport mode. Shippers of general cargo, including manufacturers of cel-lulose and fertilizers, whose production plants are located on river banks, would be more willing to shift cargo to inland shipping than the shippers with no such access.

Unitized cargo shippers pointed to the need for high quality inland waterway connections with the planned deep sea container terminal in Świnoujście, and, on the other hand, the corresponding need for development of feeder lines in the seaport of Szczecin.

An important group of shippers, who irregularly express the great need for inland shipping, includes producers of project cargo as well as river shipyards.

The results of the research carried out among the shippers were then confronted with the opinions of the forwarders. In their opinion, inland shipping which serves the hinterland of the seaports of Szczecin and Świnoujście will mainly compete with rail transport. The entities involved in the survey indicated that in the case of rail transport, time or lack of indirect handling are not the most important factors determining the choice of transport mode. Transport cost was indicated as the key factor.

The surveyed forwarders were also asked to estimate the scale of savings expected by them with regard to the costs of inland waterway transport compared to rail or road transport. 50% of the surveyed forwarders indicated a 10%–20% level of savings, while the other half of the respondents thought this should be 30%–50%. The higher thresholds indicated by some of the forwarders expressed their attachment to the logistic chains they currently used.

Over 60% of the surveyed forwarders did not provide a specific, acceptable transport time for the entire consignment. To most of the forwarders, ensuring 250 navigable days per year is enough to shift cargo towards water transport.

In response to Question 3: *"Assuming that the expected demand for transport is met, what volume of cargo would you be willing to transport?"*, taking into account the identified factors of hinterland transport competitiveness, the majority of the shippers and some entities from the "others" group (like power plants and petroleum industry companies) presented a diversified approach to the modal shift. The most frequent answer was the possibility of shifting 50% of transported cargo mass to inland shipping.

The study carried out on shippers also provided evidence that bulk cargo transported to/from the seaports, including power coal, coking coal, coke, ores, scrap, timber, chemicals and diesel oil, gravitate towards inland waterway transport to the greatest extent. The inland shipyards located on the Oder River will be able to ship several or more inland vessels a year via the seaports, and the intermodal operators empty containers in numbers that are difficult to estimate.

5   Conclusions

The recent literature acknowledges that transport cost is the most important factor e.g. [21, 22, 26]. However, our study provided evidence that the most important factor is to assure reliability of delivery. It is the necessary condition for a modal shift between competitive transport branches. Lower transport costs without the certainty of delivery on time will not convince shippers to a modal shift. Access to high quality inland waterway connections with the seaports is a prerequisite for inland shipping to meet the shippers' expectations and preferences being the factors of transport competitiveness. Consequently, increased competitiveness of inland shipping will contribute to the modal shift from road and rail, as well as it will improve the competitive position of the seaports by shifting the demand from the other transport corridors (from other competitive seaports). In the case of the OW, this condition may be met if the planned modernization investments are completed.

Our research contribution to the literature is also the hierarchization of importance of these factors (Fig. 3) and identification of dependences between type of cargo, length of transport chain and transport costs.

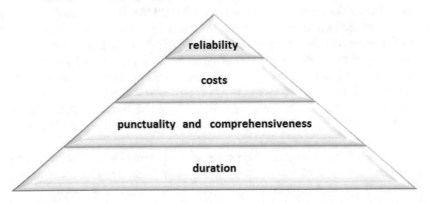

Fig. 3. Hierarchy of importance of competitiveness factors of hinterland modes of transport in land-sea transport chains (Source: own work)

The cost of transport in the entire transport chain comes as the second most important transport competitiveness factor (the sufficient condition). Our research study showed that the tendency for a modal shift from road/rail to inland shipping depends on cargo type, size of consignment and total transport chain length:

1. If the prerequisite is met, the cost of transport is in fact the only factor determining the modes of transport selected by bulk cargo shippers. For them a 10% cost reduction in relation to the ones offered by rail or road transport is enough to shift a significant part of their cargo towards inland shipping (on average 50% of the transported cargo volume). A 20% cost reduction is needed when transfer in transport nodes causes cargo losses.
2. For shippers of general cargo, the requirement of a 10% cost reduction is the sufficient condition only in case of long transport distances (usually intercontinental transport). For smaller distances the time factor is more important. Our analysis partly confirmed the research results of Frémont and Franc [11] which referred only to combined transport (modal shift is possible when prices of combined transport are 10%–20% lower than in road transport).
3. The tendency for a modal shift to inland shipping increases with the size of the cargo consignment and transport distance. The bigger cargo consignment and transport distance, the more important the cost factor is. Big companies which fulfil big intercontinental contracts are more willing to implement modal shifts. In the case of such shippers, a reduction in transport costs relating to the overall costs of buy and sell transactions is more evident. In the case of small enterprises, the cost effects of modal shifts are limited, which results in their stronger attachment to road transport.
4. The tendency for a modal shift to inland shipping is lower when the whole transport chain is organized by an external forwarder. Their profit is not directly dependent on lower transportation costs, so they are using proven, although more expensive, transport solutions.

Transport time (duration), which in the literature on the subject is of equal importance with other factors of competitiveness, including cost and safety, falls in importance in the case of land and sea transport chains. In intercontinental maritime transport, the pre-haulage time is practically of no importance at all. This trend is reflected in the policy of "slow steaming" which is operated by the main container ship owners. However, punctuality of transport, more than the absolute transport time, is more important. The shippers' requirement is that transport is carried out for a specific period of time and if this is the case they accept a longer transport time which is, however, compensated by its lower cost.

Comprehensiveness of transport also plays a significant role in handling both bulk and unitized cargo. Comprehensiveness of transport is expressed in how the whole transport process within the framework of a river-land transport chain is organized by a logistic operator. This factor is of particularly big importance in terms of decisions made by the shippers whose business operations are performed within a longer distance from the river. In this case inland shipping requires that additional actions be arranged in relation to changes in the transport modes on the land part of the transport process.

Acknowledgements. This research outcome has been achieved under the research projects MUS 8/S/IZT/2017 and US 503-2500-230 675 financed from a subsidy of the Ministry of Science and Higher Education for statutory activities.

References

1. De Langen, P.W., Chouly, A.: Hinterland access regimes in seaports. Eur. J. Transp. Infrastruct. Res. **4**(4), 361–380 (2004)
2. Notteboom, T.E., Rodrigue, J.P.: Port regionalization: towards a new phase in port development. Marit. Policy Manag. **32**(3), 297–313 (2005)
3. Van Der Horst, M.R., De Langen, P.W.: Coordination in hinterland transport chains: a major challenge for the seaport community. Marit. Econ. Logist. **10**(1–2), 108–129 (2008)
4. Panayides, P.M., Cullinane, K.: Competitive advantage in liner shipping: a review and research agenda. Int. J. Marit. Econ. **4**(3), 189–209 (2002)
5. Visser, J., Konings, R., Pielage, B.J., Wiegmans, B.: A new hinterland transport concept for the port of Rotterdam: organisational and/or technological challenges? In: Proceedings of the Transportation Research Forum. North Dakota State University (2007)
6. Frémont, A., Franc, P.: Hinterland transportation in Europe: combined transport versus road transport. J. Transp. Geogr. **18**(4), 548–556 (2010)
7. Caris, A., Macharis, C., Janssens, G.K.: Network analysis of container barge transport in the port of Antwerp by means of simulation. J. Transp. Geogr. **19**(1), 125–133 (2011)
8. Monios, J.: The role of inland terminal development in the hinterland access strategies of Spanish ports. Res. Transp. Econ. **33**(1), 59–66 (2011)
9. Brons, M., Panayotis, C.: External cost calculator for Marco Polo freight transport project proposals-call 2013 version. In: JRC Scientific and Policy Reports, JRC81002. Institute for Prospective and Technological Studies, Joint Research Centre, European Commission (2013)
10. Álvarez-SanJaime, Ó., Cantos-Sánchez, P., Moner-Colonques, R., Sempere-Monerris, J.J.: The impact on port competition of the integration of port and inland transport services. Transp. Res. Part B: Methodol. **80**, 291–302 (2015)
11. European Commission: Inland waterways. What do we want to achieve? http://ec.europa.eu/transport/modes/inland/index_en.htm. Accessed 21 June 2017
12. Kreft, A.: The concepts of development of the Oder Waterway. The cost of upgrading OW to class III and IV. The strengths and weaknesses of the proposed solutions. In: Presentation in Conference Waterways Expo in Bydgoszcz (2014). (in Polish)
13. The Journal of Laws of the Republic of Poland: The Ordinance of the Council of Ministers of 7 May 2002 on the classification of inland waterways (2002). (in Polish)
14. Data of the Szczecin and Świnoujście Seaports Authority (2015)
15. Konings, R., Ludema, M.: The competitiveness of the river–sea transport system: market perspectives on the United Kingdom-Germany corridor. J. Transp. Geogr. **8**(3), 221–228 (2000)
16. Hunt, C.E.: Thirsty planet: strategies for sustainable water management. Academic Foundation, New Delhi (2007)
17. Konings, R.: Intermodal Barge Transport: Network Design, Nodes and Competitiveness (TRAIL Thesis Series No. T2009/11, Doctoral thesis). TRAIL Research School, Delft (2009)
18. Ministry of Maritime Economy and Inland Navigation: The Assumptions of the Development Programmes for Inland Waterways in Poland for the years 2016–2020 with an outlook to the year 2030. Ministry of Maritime Economy and Inland Navigation, Warszawa (2016). (in Polish)
19. Cullinane, K., Toy, N.: Identifying influential attributes in freight route/mode choice decisions: a content analysis. Transp. Res. Part E Logist. Transp. Rev. **36**(1), 41–53 (2000)

20. Moon, D.S., Kim, D.J., Lee, E.K.: A study on competitiveness of sea transport by comparing international transport routes between Korea and EU. Asian J. Shipp. Logist. **31**(1), 1–20 (2015)
21. Danielis, R., Marcucci, E.: Attribute cut-offs in freight service selection. Transp. Res. Part E Logist. Transp. Rev. **43**(5), 506–515 (2007)
22. Crainic, T.G., Kim, K.H.: Intermodal transportation. Handb. Oper. Res. Manag. Sci. **14**, 467–537 (2007)
23. Sierpiński, G.: Travel behaviour and alternative modes of transportation. In: International Conference on Transport Systems Telematics. Springer, Berlin, pp. 86–93 (2011)
24. Liu, Z., Meng, Q., Wang, S., Sun, Z.: Global intermodal liner shipping network design. Transp. Res. Part E Logist. Transp. Rev. **61**, 28–39 (2014)
25. Arencibia, A.I., Feo-Valero, M., García-Menéndez, L., Román, C.: Modelling mode choice for freight transport using advanced choice experiments. Transp. Res. Part A Policy Pract. **75**, 252–267 (2015)
26. Lupi, M., Farina, A., Orsi, D., Pratelli, A.: The capability of Motorways of the Sea of being competitive against road transport. The case of the Italian mainland and Sicily. J. Transp. Geogr. **58**, 9–21 (2017)
27. Wiegmans, B., Konings, R. (eds.): Inland Waterway Transport: Challenges and Prospects. Routledge, New York (2016)
28. Meersman, H., Sys, C., Van de Voorde, E., Vanelslander, T.: Road pricing and port hinterland competitiveness: an application to the Hamburg–Le Havre range. Int. J. Sustain. Transp. **10**(3), 170–179 (2016)
29. Song, D.P., Lyons, A., Li, D., Sharifi, H.: Modeling port competition from a transport chain perspective. Transp. Res. Part E Logist. Transp. Rev. **87**, 75–96 (2016)
30. Kim, N.S., Van Wee, B.: The relative importance of factors that influence the break-even distance of intermodal freight transport systems. J. Transp. Geogr. **19**(4), 859–875 (2011)
31. Wiegmans, B., Konings, R.: Intermodal inland waterway transport: modelling conditions influencing its cost competitiveness. Asian J. Shipp. Logist. **31**(2), 273–294 (2015)
32. Bergantino, A.S., et al.: Taste heterogeneity and latent preferences in the choice behaviour of freight transport operators. Transp. Policy **30**, 77–91 (2013)
33. Xiao, T., Qi, X.: A two-stage supply chain with demand sensitive to price, delivery time, and reliability of delivery. Ann. Oper. Res. **241**(1–2), 475–496 (2016)

Author Index

© Springer Nature Switzerland AG 2019
G. Sierpiński (Ed.): TSTP 2018, AISC 844, pp. 265–266, 2019.
https://doi.org/10.1007/978-3-319-99477-2

Printed in the United States
By Bookmasters